RUSSELL CELYN JONES

AN INTERFERENCE OF LIGHT

VIKING

VIKING

Published by the Penguin Group
Penguin Books Ltd, 27 Wrights Lane, London w8 5tz, England
Penguin Books USA Inc., 375 Hudson Street, New York, New York 10014, USA
Penguin Books Australia Ltd, Ringwood, Victoria, Australia
Penguin Books Canada Ltd, 10 Alcorn Avenue, Toronto, Ontario, Canada m4v 3b2
Penguin Books (NZ) Ltd, 182–190 Wairau Road, Auckland 10, New Zealand

Penguin Books Ltd, Registered Offices: Harmondsworth, Middlesex, England

First published 1995
1 3 5 7 9 10 8 6 4 2

Typeset by Datix International Limited, Bungay, Suffolk
Printed in England by Clays Ltd, St Ives plc
Filmset in 11/15 pt Monophoto Baskerville

A CIP catalogue record for this book is available from the British Library

ISBN 0–670–8–4308–3

For my mother
Grace Amelia Jones

And in memory of my father
Richard Celyn Jones

23 AUGUST 1904, 52°45′N 4°37′W

On her final voyage the Bella Maria sailed from Livorno with a cargo of marble to Syria, in ballast to Cagliari, Sardinia, and is presently bound for Liverpool with cork and oil. At 0215 in the Irish Sea, with fore, main and mizzen sails all trimmed for a broad reach on a port tack, she runs into violent squalls. With winds gusting to a Force 9, the master puts the ship in irons. Lying hove-to in the dark, all hands begin reefing the sails, while the Italian mate on the aft deck attempts a fix on their position, staring at the blue mountain range for transits. But he sees what could be an eye of a buck catching the moonlight and is distracted.

The Bella Maria tumbles about in a heavy swell running from the SW and drifts on a strong flood tide running SW by S towards the lee shore. Short steep waves break over a ledge of rocks to the north. The young mate follows the track of the swells and cannot see any future beyond the middle distance. The Bella Maria has been his only home for seven years. Due back in Livorno in ten days she is to be broken up and deleted from the register book. There is nothing intrinsically wrong with her. But these days yachts can no longer compete for freight against the steamships. The Bella Maria was making her agent a loss. He knows that no steamship will ever feel the same, sound the same, as this three-mast schooner, Carvel-built, with copper bottom and a figurehead of Garibaldi tucked under the bowsprit. It is the end of a life for him.

Lying beside him on the deck is a solid koa wood surfboard, twenty feet long, that he acquired while sailing through the Polynesian Islands last January. He doesn't know exactly where he is now, is ignorant of land geography, but is sure of the moment. It

I

needs no further thought. Unseen by the rest of the crew he slips the knots on the rope securing the surfboard and lets it fall into the sea. The board spears the water, shooting back out again, briefly twisting in the air like a marlin. It floats the right way up, sulking around the hull. He follows it over the side on a rope, hand over hand, fully dressed. In the dark he kneels on the board and paddles with his hands towards the lee shore.

Moonlight casts his shadow on the water. The board lifts and falls as swells pass beneath him. He steers clear of the rocks to the north and heads for where he hears the shorebreak pound the beach. The board suddenly accelerates as a wave rushes him to this new land. Then he does another instinctual thing. He stands up, unsteadily, imitating the Hawaiians, keeping his course by dipping his foot off the side into the wave to act as a rudder. White water flies about his ears and face, with a noise like artillery, as he surfs into the unseen shore towards the mountains pencilled in by the moon. Then quite suddenly the board grounds in shingle and he is pitched on his face in the sand.

I NOVEMBER 1959

It's been a bent scurry for fifteen hundred feet over snow on loose rock, hand over hand. Hard toil, not the pilgrimage it once was as the daily climb to work. At two thousand feet tons of distilled air nourishes me back to strength. Almost. The land blazes green in places, barren and terse elsewhere without an offering to make. The gates of the quarry are rusted open. A stiff wind nudges me through and streaks ahead in passionate flurries, in solid streams of dust. As I go deeper inside, every gesture I make – a flick of the wrist, a change of grip on my knapsack – is magnified in this petrified coliseum of blue slate.

FALL 1937

The only time Paul Gravano ever lied to me my eyes filled with blood. I remember it in the way you recall getting a wound by looking at the scar tissue. The day had started fine enough with blue skies marbled with cumulus. Cold but real clean. A shadow ran fast beneath our feet and looking above I saw the air purple out. In another minute it was convulsing. The wind picked up from nothing to a hundred knots. Men came down off hemp ropes, sluicing rainwater between their fingers. Frozen rain was flung horizontally by the wind that blew over eleven-year-old apprentices. We laboured on in near darkness while less than a mile away the sun poured over neighbouring peaks.

Paul was meditating on exposed rock, his face hidden under a bowler and behind a thick beard, like a yashmak, with just his eyes unveiled. The rain seemed to part each side of him. Every few minutes he shouted up coded warnings to his brother Jacob, sitting in a coil of rope twenty feet above, leaning against the shadow of the wind. He was boring a blasting hole into the slate, hammering, twisting, hammering the drill, spraying stuttering sparks, the frost in his beard revealed in the brief gasps of light. With each impact of the hammer, shock waves travelled up his muscled arm and through his whole body. The rope pinched a little more tightly around his femur and Paul fretted a little more down below. The wind swung him back and forth and he hit the drill on the rise. I watched him intently, so much so that when he eventually lowered himself down he

4

left behind a man-sized hole in the mountainscape.

I edged closer to Paul who exuded security, offered sanctuary. Less than a month in the quarry, and already wondering if I could survive the whole winter, I sought his counsel. A stranger to me then, he heard me out without once detaching his eye from the wall soaring above us and running with rainwater. 'Is this storm a freak, sir?' I shouted. 'Is this an historical event, or what I must expect all winter?'

'After September the weather can no longer be trusted.' His smile was a grimace, held against the driving rain.

'Then it gets worse than this? I mean, it's only September.'

'We had a storm here, eight years ago I think it was. Ton blocks of slate were carried off by the wind. The tin roof came off the smithy, rose into the cloud humming the hymn "I See from Afar the Day That is Coming".'

My head was unprotected. Little needles of ice propelled by wind cut into the soft skin of my eyelids. Blood diluted by rain dripped from my eyelashes and crimsoned my vision.

Then the wind dropped, sort of sank into a hole in the ground. Black cloud was sucked away to another mountain where the sun shone just moments ago, its sulking shadow flying over scree and pasture. Grass burned like jade emerging from darkness, forests steamed, a patchy silence descended – little parcels of silence, like boxes without walls, that the bleating of sheep, the streams, the chipping of slate could not penetrate.

I searched the huge dome of sky seeking the storm's point of departure. After the gloaming so much sudden light was painful. The sun glanced off wet mirrored surfaces of blue, spar, burgundy and green slate. Smoke from the hospital incinerator rose vertically again, puffing its stench into the air. The hospital was an amputation centre. All that hanging from ropes . . . Surgeons were busy guys over there in the infirmary. Or so they informed me.

The slate exhaled a cold breath. Paul leant close so as to listen to its voice. For a considerable part of each day he stared into stone, his eyes flooding with feldspar, quartz, muscovite. For half the year those mountain peaks were covered in implacable ice; in spring the ice melted but he opened no account. Like the mountain itself, Paul was the remotest of forms. He was consumed by the judgements he made over the proclivities of slate. Any chance we had of winning valuable slabs depended on his ability to see into the future. His guess had to be pure. It was a gift that removed him from other men.

We had reached a depth where the rock was pale, as though a sickness had passed through the seam eighty million years ago. It was unfortunate, Paul said. Pursuing sick seams could be a kind of self-destruction. Men laboured for months in pursuit of what they believed was good rock deeper in, only to see the slate crumble between their fingers. He was lamenting all this to Jacob and their father Sal – a family team who wore the same coloured fear and the same coloured joy. I was not family and stood aside waiting for Paul to bring his verdict.

I watched their fear disperse when Paul announced, 'There's a health buried inside. Three foot in, maybe four.'

His work done, his prediction cast, Paul redirected his attention to Jacob, who was swinging against the face, sitting in the tourniquet-like turn of the rope, pouring an exact measure of gunpowder from an alloy conical container into the hole he had chiselled, rushing to make the 0900-hour blasting. He measured off a time fuse in lengths of his arm – three minutes per arm, from elbow to fingertips. A

6

bugler appeared at the head of the quarry and played a short sweet arpeggio. Dozens of fuses were touched simultaneously. Thousands of quarrymen swarming across their slate nests until this bugler's moment quickly made for cover: trotting like broken mustangs, kicking up mud behind their feet.

I followed the Gravanos into a crude shed on our gallery floor. It was dark as night inside and rich with smells – sweat and blood, tobacco, piety – stewed by the sheet-iron stove. The lonely howl of the steam whistle dragged its hind legs through orifices in the slate shed, the 'penthouse', behind the clank and shudder of pulleys on rusting steel cable.

A dozen men hustled for the fire, roped off from me by their language. They used nomenclature and metonym I hadn't heard – such as crych, sparry vein, milk-thumbs, braddon, metamorphic dike, alto ripple – all quarrying words lively in the winter air. I tried to start a dialogue with one or two and it fell like dust at my feet. Watering eyes stared straight through me as though I were a trick of the mountain light. It was not that they disliked or misunderstood me, it was just that their language had a demographic range of a few hundred square miles and hearing me chew it over on a mongrel's tongue confused them. When you travel all over as I had done, extraordinary things happen to the texture of your vowels. In the United States I had worked among Germans, Poles, Italians, Irish, collecting voices like souvenirs. On my first day in town here, I scattered some small children by simply asking directions to a boarding-house. They ran off like scared jackals, figuring it was some devil's glossolalia they'd heard.

Once my eyes got accustomed to the dimness I could make out Sal's outline, his shoulders sculpted into a

permanent arthritic hunch. All morning he had been helping me split great stone pillars and lever them into wagons,
then push the wagons, our bodies fully stretched, to the
water-balance.

In the awkward silence I created, Sal came out with a
helping hand. 'I come from Antioco,' he said. 'I once
outsider too.' He read my blank expression. 'Is small volcanic island by Sardinia. One time Phoenicians they live
there. They sacrifice their children on stone slabs, to Baal.'

His account was cut short by multiple explosions, seconds
between each one, cracking open the rock. The sound was
emotional, atavistic, as inner chambers were revealed to
the light for the first time in millions of years. Slate shards
flew past the doorway and showered the turquoise lake
below.

Emerging into the light I felt an involuntary claustropho-
bia as if the quarry were one small crowded room. The
explosions had displaced the parcels of silence to different
places around the mountains. Through wheeling dust I
sighted a lone shepherd above the snowline, his collar
turned up against the cold. He sat on an escarpment
scraped clean by glaciers, a witness to the leap and twist of
primitive craft, of sons, fathers and granddaddies acting out
their ancient ritual.

I was to see that shepherd again on other occasions,
moving on at dusk when the quarry closed to avoid any
contact with gregarious people. The shepherd did not care
for other men. Empty of language, he belonged more to the
mountain than a man should rightly dare, retiring at night
with his dumb animals between the dripping walls of his
stone hutch.

His sheep scuttled across the slag, yearning after the rush
of water below, making percussive noises with their feet

against the brittle stone. Above, the mountains appeared to me as ocean swells, the way they unfolded, with mist like spindrift floating between the peaks.

1 NOVEMBER 1959

From the bluff I can see grey, blue and purple waves march silently leeward, with headsets of pale spindrift. The offshore wind irons out chop and holds them up, these tired Atlantic travellers, for a few moments longer than they would otherwise. I take out my field-glasses from the knapsack and focus on the sea, bringing those pyramids up here, like a dog to heel. Framed in the circle they make clear sense. There is an order to their form.

There is someone down on the beach, up to his ankles in the sea, floating a surfboard in the water, turning it over, rubbing sand into the deck. Beyond him by a quarter of a mile a set of waves breaks over an outcrop of exposed rock. The foam backs off and the waves re-form over sandbars nearer to the shore. The displaced water sweeps the bay at adjacent angles, creating a temporary channel of calm water, a parting between the waves.

He pushes his board forward and climbs on. A hundred yards out the shorebreak regroups. A great foaming wall leaps towards him. It looks yielding, its soft contours seem sympathetic. When the wall smashes into him his head is knocked backwards by the force. He fights to cling on to his board in the turbulence.

He is faced with a line of endless rolling mountains. I think of salmon trying to swim up rapids. He makes another twenty yards before another wedge of bouncing water throws the nose of the board up into his face.

He gains and loses ground but progresses slowly. He

skids down the back of a wave a split second before it dumps its heavy load.

Clear of the sandbars, he circles around behind the rock where the line is purest. Waves break out there for the first time since beginning their voyage from the epicentre of the storm. They come chaste but harbouring a fantastic propensity for violence. He seems to appreciate this. He is running scared, respectful. I sit watching him hover on the surface, paddling parallel to the shore every few minutes, fighting with the eastbound tide.

An irregularity on the horizon gives notice of ructions soon to come. A set of three waves banks around the beachhead, each one larger than its predecessor. He lets the first one go, to be greeted by the second in the set, feathering at the lip, shivering on the brink of collapse. He slips his weight to the back of the board, spins it round and begins stroking for the shore, moving recklessly towards the rock. Through the glasses I can see red crabs clawing out of fissures.

The water is sucked up by the wave. His arms tangle in acres of kelp. The stern is lifted and the board begins to accelerate. He gets to his feet in one move, crouching to lower his centre of gravity, leans out and turns away from the rock, into the face, and begins to react to what the wave is doing.

I have the wave perfectly framed as it forms into a sheet of blue glass, scored with kelp and streaked with foam. It seems to be building itself, higher and higher, peaking at two and a half times the surfer's height. Seven feet of surfboard flaps in front of him, pushing back hard against the balls of his feet. The board shakes, insufficiently stream-lined to take this speed. The lip of the wave snaps, curls and puts a roof over his head. For a moment he vanishes

inside a tunnel, with a curtain of foam lifted off the translucent wall by the exhaust of the breaking section. From my
angle I can just see him crouching small inside the tunnel,
lit up by imprisoned sunlight, his face close to its face,
shaving the water at great speed. I hear two distinct sounds
working in counterpoint: the wave disembowelling itself
against the limestone floor and the sizzling of a million air
bubbles collapsing into smaller and smaller bubbles.

Everything about this scene excites me. I wonder now.

I wonder how that guy down there learnt to surf. His
skills remind me so much of quarrying skills, making the
guess on how a wave is likely to give, in the way he takes
fleeting readings of wind speed, swell direction, tide, current, surface texture and colour – all of which affect the
shape.

He has deliberately placed himself in maximum danger,
taking off right in front of the rock and riding out of
trouble on a turbine of growling water. There is a conflict
of tensions: the sea in a state of flux, the rock permanently
rooted to the spot. He negotiates between these two elements, between the temporary and the static, with grace,
flair and wit.

I watch him for an hour. So far he has made every ride,
a measure of the accuracy of his judgements.

Each wave he surfs is different in many minute ways.
Like the human face, no two are exactly alike. Each has
distinct moods and drift. I recollect them all, like intense
characters encountered for a short term. The waves that
make the biggest impact on me are those that yield be-
nignly, only to turn murderous a second later. With the
glasses I cut out the middle ground, bringing him and the
waves inside my head.

WINTER 1937

Snow was falling steadily everywhere but the mountain top: preternaturally apposite for holy grounds. Flurries bundled down ridges, covering scree and slag tips. It hissed against the American Steam Devil, froze the Monsieur Blondin wire.

Through this blizzard the quarry manager, Edward Manning, appeared on our gallery, swinging a silver-tipped cane, melting the snow on his black velvet cloak. An accountant by trade, Manning's job was to make the quarry more cost-effective. Appointed a few months ahead of my arrival he had yet to save his boss a single penny. He knew no more than I did about slate and the men who excavated it. Walking around, Manning earned a thousand derisive hoots and whistles from the men, who regarded his axioms of efficiency, profitability, as an English solecism. The men's principles had nothing to do with the market. The quarry was a sanctified place for them, the value of work sacramental.

Up on the gallery roof, Jacob was using a wedge to tease out a pillar. Manning stepped sideways as the thirty-foot pillar came sliding down. He removed his white gloves and made for himself an arena in our plot, kicking two short pieces of steel rail out of his way before planting the end of his cane in the ground.

Shaking his head, muttering in Italian, Sal rearranged the rails, one length jutting over the edge and in contact with the other length. The rails had been placed end to end to show up any slide in the rock face below. A safety device that Manning didn't care to appreciate.

Paul was concentrating on the face with both palms on the slate. He seemed unfazed by Manning creeping up on him. 'Have you a good reason for being so idle?' he asked. 'I've been watching you standing around in the same spot, doing nothing for an hour.'

'Mr Manning . . .' Sal was at Manning's feet, excitedly pointing up at the mountain peak. 'You see the mountain move?'

'What are you saying, man?' Manning's face was a study.

'When a mountain move like a ship, some men they feel a stirring in their hearts.'

Manning tottered and followed Sal's eyes to the top of the mountain where a band of cloud passed just below the gold peak, creating an illusion of the mountain moving, like a great rusting steel ship through a fog.

He waved Sal away, recomposing himself. 'Have you men made a guess on this bargain yet?'

Paul emerged from his trance. 'I have, Mr Manning.'

'What bounty?'

'Eight shillings a ton . . .'

'That's too high. It's good rock here.'

'Look at the section again, Mr Manning, like I have for the past hour. What do you see?'

'It's quality heather blue.'

'Can you not see the dry lace faults? And here . . . Look, a knife joint.'

'I can't see anything.' Manning sounded desperate.

Paul ran his finger over the face. 'Two feet deeper into this rock and it's going to split along a curve. Aye, and there's a length ripple as well, running into the seam.'

The face looked as pure to me as it did to Manning. Paul may have been boxing clever to get the higher poundage rate, but who knew for sure? Wages were worked out in such a way to be inversely proportional to the quality of the rock; the assumption being that winning slates from bad rock was more labour intensive. I saw now how Edward Manning got taken. They knew everything, he knew nothing. They manipulated his uncertainty into a wage-friendly bargain.

'Your predecessor, Mr Kennedy, used a magnifying glass to see the splits. Would you like us to get one for you, Mr Manning?'

'There is no dew of sentiment clouding my eyes. I see what I see.'

On the day of Manning's visit to our bargain, the same little ritual was being duplicated in every corner of the immense quarry. Family cabals everywhere conspired in their gothic stone lairs, trying to get one over on the setting agents, the secrets of the slate coiled tightly on their tongues like esoteric worms. Many men had worked the same patch of rock for two or three generations. They moved deeper in each year but never more than twenty feet across. A bargain was like an office leased by the management that you passed on to your sons and grandsons. The men who 'owned' bargains were the aristocrats of the quarry and regarded the rights of inheritance as inalienable and sacrosanct.

As an American I entertained a casual disdain for worlds so constant and unchanging. How did people grow when there was no variety of routine? The quarrymen's community had been there for two hundred years, but it seemed much older. Working with mountain stone – there was something antediluvian about the activity. Nothing changed but the bruising and diminishing of the mountain, carted off in rectangles of minimum thickness to hang in the sky halfway round the world, roofing the British Empire. No man travelled more than two thousand feet – latitudinally – in a lifetime. They had no dream of leaving, no motility. The work was always going to be there for them and for those who came after, who would be just as grindingly poor. No amount of deprivation seemed to hurt these formal protestant rock-kickers. Overcoming the hardship of quarrying was the key to their faith. The mountain ordained them.

Paul and Jacob would inherit from Sal, who dug his bargain out from behind a rockfall eighteen years earlier. In 1919 he had been given a six-month contract without wages to clear the rubble on the agreement that he could have the bargain behind it when he'd finished. Sal's agreement was with Mr Kennedy, who was just as likely to give the bargain to someone else offering the right bribe. But the risk was worth it. For Sal, an Italian immigrant, it was a promise of a new life for him and the generations of his family to come.

Sal was a man of conviction. He acted on instinct, as when he had jumped ship. The one time I asked him why he jumped ship *here* of all places, he simply said that the mountains had reminded him of Sardinia.

The quarrymen were protective of their skill because that skill was all they had to preserve themselves, to stay alive. Passing on the knowledge down blood lines was an inheritance that safeguarded the family. Every day of his working life Sal saw older men who had never won a bargain scrambling around the rockfalls for meagre wages, and he muttered prayers of gratitude.

Manning called Paul's bluff: 'Six shillings is what I am

16

offering. Accept the bounty or look for work elsewhere. There are many others who will take your job.'

'Six shillings it is then, Mr Manning.'

Manning walked away pleased with himself, the snow quickly filling his tracks. Sal clasped his hands together. 'Thank you for this very best day.' He winked at me. 'Paul, my son, he very seducing man, no? We eat steak this month. Sirloin meatball.'

At sundown a multitude descended the mountain, carving up berrybush clusters under the steady drip of trees. They followed the river rushing noisily over granite boulders, euphoric from their day's victory over slate, the most secret of all stone. Of that hook-nosed and bearded multitude, each man would know by name a thousand. This can only happen to men who live and die in one small town. For me it was a different experience, like bathing in the rain with strangers.

Through silver birches little fairy-tale houses appeared, their lights burning in the windows. In town we passed draper, chip shop, bookseller, publican, blacksmith, and peeled off down side roads by the hundred, steam rising from moleskin coats. Doors were thrown open. Witnessing our return, the publicans wiping tankards, the butchers removing bloody aprons, all made elliptical remarks. Flung sentences. It was part of the daily life to share as many greetings as possible between the phases of work and rest. These men were just blades of grass to me for a very long time.

I remember some names: Jonathan Batty and T. S. Newlin, who I watched smoking pipes while carrying out Home Office regulations to remove the metal hinges from powder boxes. The boxes were full at the time. Nest Jones, the marker, who consistently let men slip in up to twenty-five minutes late. Jonas Mount, Peter Reading and David Wyn, three badrockmen, who stole slates to repair their roofs. Nye Adams, the setting agent, who accepted in the space of one week two exotic cats, a crate of rhubarb wine, a chicken and some onyx jewellery.

There were guidelines on how to behave in the presence of bribers. It was pretty risible stuff. Walking with the Gravanos downtown one time, I saw them place their hands to their lips in a synchronized gesture as we passed two such greasers outside the bakery.

Sharon was an amusing town. It had a kind of childlike appeal, as in a Hans Christian Andersen fairy tale.

Sal Gravano took me in as a lodger. He remembered what it felt like to be deracinated and rescued me from the boarding-house of the miserly Mrs Prothero who'd bang on my door as I lay in bed reading before going to sleep. 'Mr Lewis,' she'd complain from the other side of the door, 'you've got the lamp on in there.'

Sal's house stood in an isolated terrace on a hill above the town of Sharon. The land swept at a steep gradient under the terrace, bottoming out two hundred feet below, banking up again for a non-stop ascent of three and a half thousand feet. Visible from the house was the quarry, a purple cicatrix in the mountain flank. A stream slid past the terrace in a foaming misty torrent pouring into the river below. A curtain of cross-weaved chains stretched across the stream to stop sheep being swept away.

The entire terrace was built with slate, like everything else round there, not just the houses but the fence that corralled the terrace to keep the wild critters off the vegetable plots; slate posts sewn together with wire that I searched for sparry veins, milky thumbs and back ripple – succeeding only in parodying Paul's gifts. Even the street was named after one of the galleries in the quarry: Red Lion Road. A slate town, through and through.

On my first entrance into the Gravanos', I walked in last. Our hats came off as we passed over the threshold. The air was sweetly scented with burning elm. We hung our coats on wooden pegs behind the door, a door without a lock or key. Inside the ceiling was so low my head brushed against it. On the table a small banquet was laid out like a Thanksgiving treat. A ham roasted in cherries, a round of mustard cheddar, homemade fruit cake, butter-milk – arranged on white damask. A varnished maple tea trolley was creaking under the weight of a drum-sized pewter teapot. There was a solitary wedding daguerreotype in a gilt frame and a crucifix livening up the whitewashed walls.

Sal's wife, Rebekah, standing like a sentry from the time we came in, bowed her head on seeing a stranger enter. Her profile was manly: crow's nose, wide down-turned mouth, boxed chin – a British Pocahontas in a salmon-pink blouse, Fall-leaf skirt, white ankle socks and black hobnail boots. Sal must have still had sea water in his eyes when he saw her. Or maybe all women are handsome to a wetback with no place to go.

She stared at me out of the corner of one eye until Paul introduced us. 'Ma, this is Aaron Lewis, an American. He's going to lodge with us a while, so would you make a little room for him at the table?'

From the oak dresser she chose a fine china plate with a flower pattern embossed around its edge and laid a place for me beside Paul. The stairs came straight into the room

and Rebekah had to back up a few steps to let us through
to the table. I struck a perfect upright posture from the
moment I sat down and maintained it as a young pretty
girl of around nineteen, with a baby not more than three
months old in her arms, descended the stairs.

Jacob introduced his sister Leah. Whoever knocked her
up, they weren't saying. I avoided eye contact with the girl,
avoiding trouble. She looked as if she'd seen enough already. Rebekah held the baby so she could take her place
at the table and then handed him back to her the moment
he started to cry.

Paul asked me if I'd like to say grace. 'Sure, I don't
mind.' I cleared my throat. 'We thank you Lord and ask
that we may be called upon to remember that the wheat that
makes the bread would not grow without your light, the
meat we eat once grazed on the fields that you green in
your inimitable way . . . there would be no food nor nothing
else just desert. Amen.'

Rebekah stayed on her feet as we ate, carving a white
loaf clamped in her armpit. She cut paper-thin slices as our
needs dictated. Such self-abnegating women had passed on
in the States some time ago. Now they have museums built
around them.

But I was misled by first impressions. Rebekah turned
out to be her own woman. 'How did Manning sport himself
today?' she asked flatly.

'He thinks we're megalomaniacs. Subversive, he says.
Vain.'

She tried to cut the next slice of bread and it fell apart in
her hands. 'So the man in the black velvet coat and white
gloves thinks you're vain, does he?' She grew ever more
excited. 'What's an accountant know about quarrying?
You could be extracting teeth for all he cares. He's left a

comfy office job in London for a quarry in the ice, looking for what? I'll tell you . . . looking for experience that makes a man. While we have to suffer meanwhile. He's a young turk, that Manning. You can't acquire character like a wardrobe of fine clothes.'

'I don't think we're suffering exactly, Ma.'

'He's a young turk.'

The guys seemed embarrassed about their lack of authority over Rebekah, while Leah looked kind of proud of her mother. Jacob motioned for more bread in an attempt to dampen her enthusiasm. The peace held for a moment while Rebekah carved from the hip. She cut the bread at a dangerous angle, the knife coming to the end of its journey each time with a shudder under her fleshy arm. She lay four slices of bread on a plate before starting up again. 'Profit is all he can think about. We put people first round here, Mr Lewis. Profit has no loyalty to anyone.'

'I agree with you,' was all I could manage.

'Well, I tell you, if I was in that quarry I'd invite him home for tea. Teach him some manners. Then again, I'd probably hit him. I wouldn't be able to control myself.'

'Slice this ham, Rebekah, and to hell with him, that man,' Sal said.

Over time, I saw that Sal and his sons gave Rebekah all their wages. She acted as paymaster in the home. 'I think if a man gives the woman of the house his wages she shows a greater interest,' she explained. 'I can keep table in my own way.'

I also saw that Rebekah was prone to vanishing, often for whole days at a stretch, into a darkened bedroom. Then the house became staggeringly silent as if she were dead. Paul explained that ever since they were children they had learnt to regard their mother's sick depressions as a walled garden, as inaccessible as the quarry owner Lord Elusen's orchards. It was Leah who took over as housekeeper, growing up in an instant from child to woman and back again when Rebekah emerged from her sojourn in the bedroom, shivering with spent hallucinations. Nobody said anything. They just continued reading their library books and holy texts or cleaned their boots while staring into the blue night through the back door. I stared at Rebekah, emerging from misery into light for the first time in twenty-four hours or so. The struggle was etched into her face, her blood rummaging just below the surface of her skin, her hair pulled back tightly as if a hurricane had ripped through the bedroom. She'd pick up on some chore left behind a day before and whistle a single note of exorcism.

Sometimes, when working in the quarry, I'd picture Rebekah wandering around town, in an endless round of greetings with other wives and mothers, reminded of their dependency on the quarry by the chorus of men's voices clashing in the mountain air, in every explosion, by the horn and the siren, by the rattle of trucks on steel rails, the avalanches of slag and the occasional splash of some Icarus falling into the quarry lake after trying too hard to listen to the voice of the rock.

And I used to think of her every time we opened our snap tins. She packed a great lunch, each piece of bread and cake carefully wrapped in greased paper like Hanukkah presents. It was to her that I gave thanks when the others said grace in the noonday twilight of the quarry penthouse.

After grace we pooled our lunch buckets: buttermilk, pancakes, *barabrith*, cheese, corned beef. Wages were so varied that redistributing food was an opportunity to level things out. From the centre of the room the sheet-iron stove generated a heat too weak to dry out our clothes. A prehistoric coldness seeped up from the flagstones through the soles of our boots. We distracted ourselves from such discomforts with Bible knowledge quizzes, mental arithmetic tests, competitions to read aloud unpunctuated prose, singing contests. Sal always won the singing contests with arias from *Rigoletto*, *Il Trovatore*, *La Traviata*, *Otello*, in a fruity counter-tenor voice. On more than one occasion, Jacob gave the same lecture on nineteenth-century inventions: 'The mixed-flow water turbine was invented in 1855, making way for Davy's differential valve gear for non-rotative pumping engines in 1871. A few years later saw the multi-stage turbine pump. In 1879 the first hydro-electric generating plant was built in Hull . . .', sounding just like my old history teacher in junior high. But I am missing the point here . . . The truth was, these men liked to talk not so much in appreciation of the content but to hear the sound of their voices. Their language was rich and musical.

'Suet pudding is apparently not popular in the asylum. According to the Lunacy Commissioners.'

'Dr Williams's Pink Pills for Pale People are a cure for St Vitus's Dance.'

'Queen Victoria survived three attempted assassinations – by Edward Oxford, John Francis and a hunchback. They all tried to shoot her.'

'President Garfield was shot. He recovered but Lincoln was not so lucky. John Wilkes Booth shot him dead.'

Frequently they strayed on to the subject of war:

'The Bible says "Whoso sheddeth man's blood by man shall his blood be shed."'

'"He who lives by the sword shall die by the sword."'

'We know from Cain and Abel that killing is a natural tendency. That is why war must be avoided at all costs.

War raises unanswerable questions about the nature of faith.'

'The devil had his day in 1914 and we still haven't recovered from it.'

It was prudent to keep my mouth shut. I wasn't so crazy about war or anything, but I'd sure enjoyed my brief stint with the US Cavalry. I sat with my back straight against the wall, exhaling through my nose to keep the noise down. In the Cav I'd met cowboys from the Goodnight Loving and the Chisholm trails, along with college boys from back east. We all got on famously and I missed their company for a long time afterwards. Quarrymen shared something like their comradeship, born out of dangerous work. They too were mortal beings on an unstable earth.

Paul went down to the lake to get water to make tea. The men used the same lake to drown their cats. The tea killed more quarrymen than falling rock. If pneumoconiosis or silicosis didn't get them first. The lake was a turquoise phenomenon, whether pigmented or reflecting a synthesis of the colours of slate no one knew for sure, but the water looked clear enough inside Paul's billycan.

Paul put the can on the stove as a conversation about the South African war was heating up. Some of the old men could still remember living through it. 'There was this volunteer's wife. She was sentenced to three months for neglecting her children. Her husband was serving in the Royal Welch Fusiliers up at the front. Everyone was indulging themselves with accounts on the front page about De Wet and his Boers being chased into the Orange River Colony, while she turned to drink in anticipation of her husband coming home in a casket. She let her children go without food and fire. And all she said in court in her

24

defence was, "Oh, I suppose it's all true." Just like that. "Oh, I suppose it's all true"!'

Paul shook his head slowly. It troubled him that someone could drift so far from the constructs of civilized life. He felt personally aggrieved by that account from over thirty years ago.

We were fifteen men sitting around that table, our names nigh on covering all the Old Testament prophets. (What were our parents playing at? Did they think we would live up to such names?) One of the old guys there, a smithy called Ishmael, or was it Joshua? . . . Whatever. Anyway, he used to work the whole year through without a shirt, I remember that much, his entire body covered in hair like an ape's. He also had cataracts, a secret the men kept along with all their other secrets. This Ishmael came in on the discussion with a story about a steam packet, *Lady Elizabeth*, that was carrying arms for the German colonists in East Africa. On her return voyage she was blown six hundred miles off course by easterly trades. A third of the crew starved to death, another third died from African fever. When the ship docked at Liverpool, one of the survivors went ashore and spent the night with a woman named Leslie Ann McGregor. Some time during the night he strangled her with her own salmon-pink silk scarf. 'All that he suffered at sea came back to him in a blinding flash,' said Ishmael. 'He hangs tomorrow morning.'

'That's a terrible story, Ishmael . . .'

'The ship nearly foundered in a hurricane. Two of the crew clambered on to a ledge of rocks. They were washed away and drowned. The vessel was a very good-looking vessel too, by all accounts.

'So after he's strangled her, the sailor dressed up in her underwear.' Ishmael railroaded his way back in. 'He was walking around Toxteth in suspenders and pink knickers when the police found him.'

A sceptical frown passed down the line. It was left to Paul to try to catch him out. 'I think I saw the sketch that went with that story, Ishmael. In the *Daily Werin*. Didn't the van taking the sailor to prison break down? Yes. The artist must have been sitting there quietly painting the landscape at the time. The sailor's manacled to the wheel of the van, wearing yellow bloomers.'

Ishmael jerked his head from side to side, absorbing our smiles through his bleached and useless eyes. 'It was *pink* knickers I said he was wearing.'

We finished supper and cleaned up ready to go into town. Rebekah cleared the table as we washed the slate dust off our bodies round the back of the house, from buckets that Leah topped up with hot water. The freezing air numbed the skin on our bare backs. We laughed as though the cold were a joke of our own invention. I soaped Paul's back and Jacob soaped me, the three of us forming a happy circuit, bringing warmth to our fingers through friction. Paul had little moles on his shoulders like currants or chocolate drops that I was almost tempted to taste. I snatched hold of his waist and scrubbed his neck. Jacob was ineffectual at this. He couldn't work up a lather. I squeezed a sponge down Paul's neck, down his pants, making him shriek like a girl. He tried to escape but I was relentless until we both collapsed on the ground one on top of the other.

It was around about this time that they usually asked me about my past. We had just started out from the house walking down the hill to the town below.

'I was a breaker boy in Scranton from the age of eleven.'

'A what?'

'That's what I did, what breakers do, pick the slate out of the yield of anthracite. At seventeen I went below. Working down mines is no picnic, I tell you true, guys. It's all wet and dark and dusty. Voices drown in smoke and you get an early warning of what death is going to be like . . . Well, anyway, I left. My ma and pa came from round here, as you know. With their blessing I crossed the lake.'

'Is much a man leaves behind when he go from his country,' Sal said solemnly. 'It takes very much time before the old land to die and new roots to come and grow inside you and the new land make good fortune.'

'So far I like what I see here. Give me my own bargain some day . . .'

'And a wife. Nice local girl.'

I slapped Paul across the back. 'How come you ain't got a girl, Paul? You too skinny to be a big draw?'

'I've got work to do in the quarry. This body is just fine for that.'

'You don't do any work. Manning says so.'

'Now you mention it, Aaron, you look a little like Manning.'

'Oh yeah, oh yeah . . .'

'I met once a very seducing man and I say, "Friend, what have you to do that is so close to the pretty with the women?" And he say, "Na–thing, I just wisdom. Women they like wisdom not beefcake." I say, you all are in the age of marriage. You should listen to Salvatore.'

The brothers and I looked at each other with suppressed smiles and joined the crowds filling the High Street.

I NOVEMBER 1959

It is still early morning as I return from the bluff with ozone in my eyes and salt-blasted hair. Sharon – what is left of it – has not yet come to life. I glide past the windows of the draper, the greengrocer, covered inside with newspapers browned from the sun. Headlines and titles pitch a riddle back to me: MY BOY. *Titbits*. COME BACK, KENNY. *The People*. IN THE SOUP. SWELL GUY. *News of the World*. The only two stores trading in the High Street are the post office – run by the same Mrs Raleigh, now so crippled with arthritis that she has to sleep in the shop, with customers who serve themselves and leave the money on her pillow – and a new cinder block, A & P type of place, painted yellow and blue with a forecourt of broken concrete. Black shelving clouds loom behind like a curtain hung between two mountains. The railway viaduct passes close by and trundles lugubriously out of town. Under the arches, partially obscured by weeds, are old trucks and rusting automobile engines.

Half of all the houses in Sharon stand empty, inviting anyone to live there for free, while under the great tips of badrock a non-indigenous people live in caravans and converted coaches. Their caravans are better accommodation than the old slate stone houses, which are too old to be repaired, too solid to fall. And too ghosted. The caravan dwellers are beggars who survive on gifts and fight among themselves over those gifts. They possess none of the instincts of the people who built the original town. Their washing

28

lines stretch between the caravans. White shirts flapping in the breeze against the loose wall of poor slate fills me with a curious temper.

There are four cathedral-sized chapels in Sharon: Baptist, Presbyterian, Independent and Methodist. And a shunned and forlorn Anglican church. The vicar used to be a monoglot Englishman on the quarry payroll and few worshipped there. Chapel was safer than church. You could say what you liked without fear of recrimination. None of Lord Elusen's spies could understand a word. Or so they believed.

Inside Bethesda Independent chapel are cars being repaired. Sparks fly through the open door, the light reflecting in pools of oil and against the chrome spanners hanging on the wall.

Jerusalem is the only chapel still functioning as a place of worship rather than as a garage, a public toilet. The interior walls are crumbling with rising damp and the lead in the windows has been stripped. The only thing that keeps the pitch pine pews from walking are the six-inch nails holding them to the floor. Sharp-edged pools of sunlight hit the footworn flags; light that pulls like gravity on the eye.

WINTER 1937

Nights in Sharon had their own complexities. Time was pursued relentlessly. I followed the Gravanos into Jerusalem chapel where three hundred men and women were sitting in an oak-panelled saloon. The minister, Mr Parry, stood at the pulpit delivering a fire and brimstone sermon on Melville. Yeah, that's correct: Herman Melville. Which was hard for me to square with what the annotating eye figured was a poor and motley crowd in bleached shirts and worn grey flannels; scrubbed hands and faces releasing strong scents of carbolic.

Mr Parry raised his short arms each side of his head, careful not to displace his hair. That hair was a character in its own right. Bald, apart from tufts above the ear, Parry had grown his crop as long as a woman's and wrapped it around his head like a turban. He glued it down with squirrel fat that not even the strongest mountain wind could displace. At night he pulled out all his grey hairs with tweezers and burned them ceremoniously in an oyster shell. A man I'd once watched in the graveyard making three small children cry. 'Weep!' he shouted at them, 'as if the day of judgement has dawned.'

'The narrator of "Bartleby the Scrivener" craves security,' Parry said. 'That is what is wrong with him. His chambers are a retreat, he makes his devotions there. There are no women in the office to tempt him. Now then, what else? For thirty years he has been a scribe. The profession of the Pharisees. He appoints Bartleby because he looks safe.

How is Bartleby safe? How is he described?'

'As a man of sorrows,' someone shouted from the gallery.

'And where's that from, like? Do you recognize an echo from *the* text?'

'Job . . . Must be Job.'

'Isaiah, actually.'

The chapel was noisy with speculation. The atmosphere seemed theatrical, tense. My head jerked from side to side, trying to identify contributors, but the ball kept moving around too fast. I seemed to be the only one there who hadn't read a word of Melville. I had been to college, which meant I could behave sensitively and not let myself down in social settings. But I didn't have the kind of handle on obscure literary texts that they had. No college professor ever taught me how to apply literary theory to life like Mr Parry. In my time I've worked among construction gangs, with underwater steam fitters, locomotive firemen, coal miners, iron workers, none of whom knew what a book looked like. Nor did you expect them to. Men who have never been to school, who start their working lives at eleven, have no need to read anything but their name on a pay slip at the end of the week.

'Isaiah fifty-three, verse three . . . "He is despised and rejected of men; a *man* of *sorrows*, and acquainted with grief." Isaiah's premonition of the Messiah . . . That's how Melville describes Bartleby. Now then, what is going on here? Where are we being led?'

'I believe the narrator is a Christian, Mr Parry, let's be fair. He upholds the values of charity. On Trinity Sunday he goes to church . . .'

'All right. That is as maybe. But to do what?'

'To hear a celebrated speaker.'

'Yea! To *hear a celebrated speaker*! The church offers

31

salvation in the absence of Christ, but this narrator is going there to be *entertained*. In the event he doesn't get to church. Isn't that right? He goes into the office on the way and is sidetracked by Bartleby, who he discovers living there. So what's he do then? I think we can say he overcomes his Old Testament anger, but falls short of heeding Christ's counsel, uttered at the Last Supper, amid an atmosphere of treachery and corruption, after Judas had just left on his infamous quest . . .' Parry licked his fingers and raked through the pages of a Bible. He read from John thirteen, verse thirty-four: '"A new commandment I give unto you, That ye love one another; as I have loved you . . ."' He smiled slyly as he approached the point of his provocative hypothesis. 'The narrator denies Bartleby three times. Rejects him in favour of interests in the material world. Thrown out of the office, Bartleby is arrested for vagrancy and taken to the gaol in New York, America, arm in arm with a policeman. He is paraded through roaring thoroughfares at noon, as Christ was paraded through crowded streets to Golgotha. Why is he allowed to die?'

The silence in church was as expectant as the moment before a killing. The answer to his question lay just out of reach, with everybody working up quite a sweat in the meantime. The minister looked around, trying to catch any backslider sitting on the edge of his pew on the verge of losing faith. In the long dark seconds collapsing around us I could see Melville's grinning face on Mr Parry's narrow shoulders, his long pink tongue hanging out. 'Why does a benevolent, all-powerful God allow man, His most cherished creation, to die? If God is merciful why does He abandon us to extraordinary suffering?' He was shouting now. 'Why does the narrator, this *Christian* narrator, not save him? Why does Bartleby not save himself?'

Paul started making heat in the seat beside me. This intense young man with a Mennonite beard stood up clutching the back of a pew. 'Christ's quest is for love, in the name of love,' he said. 'It is a concern for the welfare of other people. Bartleby's quest is of extraordinary selfishness. Christ died on the Cross offering everlasting life. Bartleby fails in death to give any reassurances.'

'But why not save *himself?*' The minister stayed combative.

'Because he never received the good news,' Paul insisted. 'His previous job was as a clerk in a dead letter office. A place filled to the rafters with what Melville calls "good tidings that never reached those who die".'

'Yes. Yes. Good. Go on, Paul.'

From all the approbatory murmurs, I could tell the crowd were rooting for Paul to pull them away from the edge where the minister had led them. This might have been a game, but the stakes were high. Faith had to be exercised like a horse, tested on new ground and on old. Faith needed rough as well as tender handling to keep in shape.

'Bartleby's job at the dead letter office is referred to as his "original source". The Old Testament is an original source. The Gospels mean Good Tidings. These are clues Melville leaves for us. He gives Bartleby Christ potential, but in order to expose Christ's offer of everlasting life as phoney. But God does not save Bartleby as he did not save His own son, not because He's malevolent or powerless, but because He's just not an authority figure. Evil is the shadow in a beautiful picture, not the picture itself.'

Not a bad performance from a young man whose hands had been calloused since the age of nine. In any town in America Paul Gravano could have been somebody, a head

33

of industry, a politician. But there was no way up for a man like him in British society, only latitudinally across the pages of books. But at least he had that. When he sat down everyone applauded. He had got them off the hook, yanked them back from the big drop. He lowered his head modestly, burning with intrinsic rewards.

Paul noticed Sal sitting on the edge of things and tried to draw him in. Difficulties with the language isolated Sal. He understood everything that was said, but was too slow to compete against the verbal dexterity of the men. Paul told him that Herman Melville had once sailed with several packets, including the *St Lawrence* that docked in Liverpool in 1839. 'He jumped ship once, too. In the Marquesas Islands, French Polynesia.'

This chapel was really something, built by the people in their free time, using their own resources. They even had to pay for the slate on the roof. They tithed a portion of their wages to pay the minister who ran the chapel as a social club, an extramural college, a venue for union meetings, as well as a place of religion.

We walked from Jerusalem to the Blue Boar for a nightcap. Back home in Penn., I was used to seeing miners come up from galleries below the earth and seek fierce entertainments above. Kids as young as fourteen ran around Carbon County carrying revolvers and terrifying women leaving their knitting bees, while their fathers crawled around the grogshops, ginmills, whisky holes, rum cellars; going from prizefights – some lasting a hundred rounds – to dogfights, cockfights, pitch and toss, quoits, as the spirit waned.

The Blue Boar was no such place. We claimed a table in one of the small rooms attached to a central circular bar as Leah joined a couple of girlfriends in the lounge. In other rooms men played draughts and dominoes. The mediator was the barman, orchestrating conversations between men and women in different places. When someone began a song in one room and the chorus was taken up by men in others, the barman kept the ground beat going, working the pumps like a church organ.

Paul wanted to discuss American literature with me. Said he liked Hawthorne, Twain and Sherwood Anderson. I hadn't read any of them, I had to confess, so we let Jacob switch subjects. Jacob preferred what he called 'the world that you can see' to literature: science and technology. When he spoke his voice ran away with him as though it was always the next sentence with which he was going to make his point. 'I wish I could ... the last century. Can you imagine? Golden age for, Brunel built his ship. I wish I could have seen the *Great Eastern* on the Thames. Men riveting inside day and night. People came – it had storage for coal big enough to reach Australia non-stop – and watched it take shape in the water.'

'Yeah, I know about that,' I cut in. 'The Suez Canal was dug around then and Brunel's ship was too wide to pass through. It ended up as a floating circus.'

'It ruin Brunel,' Sal added, 'that ship.'

'If the men had taken a pay cut it might have been a different story,' I said.

'Why should the men have taken a pay cut?' Jacob asked, his roaming gaze stopping just short of eye contact.

'For the sake of pride. It was their ship too.'

Jacob suddenly became more lucid. 'Do you think for one moment Brunel would have shared his profits with them had they worked for less?'

'He might have.'

'Has Elusen ever shared his profits with us?'

35

'Would you want to share his profits, Jacob? You think he's heinous.'

'Would you?'

'There no such thing as wages of sin. Money is money. It's only what you do for it that is good or bad.'

'We're losing the point here.' Paul tried to calm tempers. 'We were talking about Brunel's ship.'

I quickly took stock and saw how far out on a limb I'd gotten. Jacob had raised the Cain in me and I had to really flap and swim upstream to find my way back in. 'When they scrapped the *Great Eastern* two skeletons were found trapped between the panelling and the hull. People went around saying it was doomed from the start.'

Jacob lowered his head and clenched his fists on his lap. 'I still would have liked to see it being built.' He followed this with a long thoughtful swallow of ale, his eyes clouding over as he travelled into himself. He came back shortly and empty-handed. 'What about the Forth bridge? Wouldn't you like to have been the first to cross it in 1889?'

Jacob had inherited his mother's appearance – high, arched eyebrows, receding hairline – and his father's personality. I could easily imagine him taking life-changing decisions on the heel as Sal once did. It is always easier to be fearless when imagination is absent. Maybe this is why women found Jacob dull. There was nothing they could get a purchase on, no foxiness or quixotic restlessness transmuting into sexual passion, while men liked Jacob precisely for his unreflectiveness. They stepped out of his way when he seemed to be on a resolute track. I caught myself wondering if he knew how to fight. Although this was not a pugilist's town he had plenty of muscle on ice if he needed it.

Jacob was the faultless splitter of slate, but one who could not judge the stone. That took a romantic, even melancholic, disposition and Paul had inherited from his mother to do just that. He was susceptible to similar depressions that wreaked havoc in her life. Except he didn't lie in bed staring at the wall but stared into the smooth face of slate instead. He had an outlet for his melancholia. He once said to me, 'I judge the rock because I can't write poetry.'

I went to stand at the bar with four empty glasses. In the lounge I watched Leah with her two friends, Barbara and Kate, sitting in the window-seat drinking brandies, watching the men shimmy up their feathers. Who did she take after? Leah was an anachronism in that family and in that protestant town. Not only had she sidetracked a wedding altogether and got knocked up, but was on heat most of the time.

The dude to catch her eye tonight was Saul P. Howells, a member of the quarrymen's committee, drinking and joking with his buddies over there. I wondered why it is that certain men are so incapable of being serious when they get together. If they need to raise weighty issues, a woman always has to be present.

Once, when I was in the house, Leah brought Howells home to meet the folks. More precisely to meet Paul. Normally it was a father who watched out for his daughters. Husband-fancying was a father's job because he would have known all the boys from birth. He'd have the chance to see if they'd grown up with backbone by observing them in the quarry. He culled the herd of apprentices and brought home the best crop to meet his daughters. There were mismatches, obviously. People change in the course of their lives. But there is something to be said for marrying a man your father saw in his pram and who you went to school with. If you knew what he was like then, there is less

37

chance of being surprised by anything he might do in the future.

She had brought Howells home from the pub, calling in the chip shop on the way. She emptied the chips on a communal plate, but Paul refused to eat. From the shadow-land of his fireside armchair he silently condemned Howells's aimless chatter, his pub talk, his easy-speak. Paul's wrecking thoughts spoiled the romance for his sister. While Howells played with the baby on the rug, Leah ducked under the flock of panicked seagulls flying out of Paul's head.

With hindsight Paul's instincts were about right. Howells just wanted all the girls all the time. And when you want everything, you end up with nothing. Nothing but a hang-over in the morning. Or Leah as a last resort at closing time.

So, anyway, Barbara and Kate were talking about the minutiae of their day from what I could hear. Leah had stopped listening to them for quite some time, to watch Howells drink a whole pint of beer in one, probably wonder-ing if he was as good as it gets. All men paled beside her brother Paul. No youth had his shaking passion.

All other activity faded out of the corners of my eyes, leaving just Leah watching Howells, stranded for a moment on his own with a fresh pint in his hand, his rosy lips parted in a smile, ready to receive the ale. Over the rim of his glass his bloodshot eyes roamed the room. As closing time was approaching he began glancing in Leah's direction every few seconds. I wondered what he would do if he looked across and she had gone. Would he leave the pub and go look for her in the street? But if Howells was ever going to follow her out of a pub he would have done so long ago. The truth was, Howells would have to be chased.

I knew the kind of guy he was.

Howells lived with his mother and father. His mother
had cooked every meal for him since the day he was born.
Howells didn't raise a finger at home; he would make a
mess of boiling water. The family ate dinner at five thirty at
the latest, so as not to ruin his drinking time, which began
at eight. Some nights in the pub were better than others for
Howells. Some nights he had not properly digested his
dinner and the beer would make him bloated. On such
evenings, even talking was too much to do. He would
return home and his mother and father would ask their
statutory question whether he'd had a good evening at the
Blue Boar, and he'd reply with the same answer that there
hadn't been much he felt like talking about. As a rule,
Monday night was usually crassly dull for Howells. Tuesday
not much of an improvement. Wednesday was middling,
whereas Thursday had its moments, as people began to
sense the end of the week. On Thursday some of the boys
got mildly drunk and laughed at each other's short jokes.
Howells would laugh at more or less anything, but got a
little bullish if he wasn't given enough time to deliver his
punch-lines. On such nights when the conversation went
awry, he'd suddenly notice how the beer was clouded, how
ugly the women were.

On Friday and Saturday nights Howells was a different
man. There was something exciting about these nights for
obvious reasons and he used to get high on account of all
the women appearing. Because he was high he would get
drunk, leaning on his buddies, laughing in their ears.
They'd never move from the bar, because that way they
could get served quicker and because it afforded a better
view of what was going on. They felt central to the general
toing and froing at the bar, since everyone had to pass
by them to get served. Standing also helped the beer slip

down. Standing in a pub was Howells's only recreational exercise. He never took off his coat, because his tobacco and money were in the pockets, and he hardly ever slipped off to use the gents in case he missed something. His bladder must have been remarkably strong. As the separate camps of men and women gradually began merging near closing time, he would usually make his move. The mixed crowd and the buzz of sex within it excited him. But he was often too drunk by this time to be any more than a slob sliding down a woman's sleeve.

That night I watched him do something unusual. He walked across to Leah's table and *sat down* with the three girls. Barbara and Kate took it in their stride, but they hadn't been studying his movements as closely as Leah (or I) had. She seemed so astonished that he gave her a look of solicitude. 'How's you?' he asked.

'Are you really sitting down, Saul, or did you fall?'

'I've come to see you lovely girls.' These were the odds Howells liked best: three-to-one. 'What you all drinking?'

'You buying, Saul?'

'I might change my mind in a minute.'

'Mine's a brandy,' said Kate.

'Make that two,' added Barbara.

'Bloody hell, I'll need a raise at work to drink with you. What about Leah?' he asked softly, generously.

She could not answer. A drink didn't seem the point. He was trawling her face with his brown eyes and she must have felt such a pressure at the back of her head, trying to force them together. She hadn't worked out her order yet, but it was going to be expensive. She whispered in his ear so the others wouldn't hear. 'I don't want anything, Saul. Walk me home tonight and I'll put in my order then.'

He nodded, the thing was fixed. As far as she was concerned he could go off now and flirt with any number of women, get drunk, make his friends laugh. They had made their contract. As he brought Kate and Barbara's brandies over in a two-hand clutch, he winked at Leah, ratifying that contract. He didn't sit down with them again but returned to the bar. She gobbled down a few packets of potato chips, keeping company with Barbara and Kate only in a marginal way. They began singing out of tune, their hands cupped around drinks, two blondes rolling on a bench, getting older by the hour.

I took our own refills back to the table. It was to be our last glass. When Sal stood to leave his sons followed him from the table. I was just getting into my stride but followed their rules of moderation and left with them.

We swung out of the pub into a sky of clear cold stars, stars that were among us, tumbling down off the mountain rim. Leah was already outside, facing down Barbara on the issue of Howells walking her home. Barbara had invested something in him herself a short spell ago, something that came to nothing. 'He asked me tonight,' I heard Leah say. 'You have him tomorrow.'

I watched Howells wean himself from his friends, and, spitting into the gutter, edge backwards to meet Leah. Only those who cared would have twigged the game plan. He walked slightly ahead of her, so as to suggest to any other female that he was on his own. We followed them for a few blocks past the crowds outside the Victoria, the Castle, the Quarrymen's Arms. Then when Paul's attention was elsewhere she and Howells cut down off the High Street and disappeared along the railway tracks.

On our way home we called into the chip shop, a veritable seafood museum. Various breeds of Koi swam listlessly in an aquarium beneath the counter in which an array of cod and haddock was piled on a hot perforated tin rack. There was a mural painted on the back wall of sharks, swordfish and rays nosing around pink coral. We splashed our order of cod and fries with malt vinegar and salt and snapped newspaper around to keep in the heat. A crowd of teenagers came in behind, inhaling the tobacco smoke and beer on our clothes, imagining the pub we'd come from to be more fun than the chapel societies they'd just quit.

We stepped out of the chip shop straight into a living stream. Shoals of men and women were walking home in the dark. Jacob and Paul kept stopping to swap platitudes and that short walk took so long I felt my patience fraying at the edges. We plodded up the hill to Red Lion Road, plucking steaming white flesh from our hot vinegary bundles. Household lamps were turned down as though by our influence, and each section of the river grew darker.

Before going indoors for the night all four of us lined up in a row, unbuttoned and took a leak in the river. Sal held his head up, taking nautical readings off the stars, and began an account not even his sons had heard before. I like to think it was my presence that helped him along; it takes a stranger to prevent a family from ossifying. He told us how the years at sea flushed out his Catholic faith. The sea, he said, is a dynamic world while faith is static. The sea destroys belief in anything but physical survival. It has no heart, no conscience, hosts no church. Prayers go unheard. Enchanting little atolls seen at a distance and proffering sanctuary to weary sailors turn into cemeteries the moment one steps ashore; fissured volcanic chapters in the sea's history where only skua gulls thrive, dining on luckless penguins created by the Lord with no means of defence.

I took a check down the line, saw how Sal had undone his belt buckle and let his pants drop to the knee. He held on with one hand while the other conducted his story. Jacob had both hands on his head and was firing at an exposed boulder in the river. Paul held his underpants open with two fingers, his dick in the other hand. He was first to finish and gave himself a rigorous shaking before firming up his underpants, imposing discipline on the region.

Some time in the closing days of the last century Sal went on shore leave in Oahu, Hawaii. Always drawn by heights, he started out for the Waianae mountains while his shipmates raided the villages for women. At the top of the mountains he discovered a church whose roof had been shorn off in a storm. Its walls were crumbling with termites and inside an American Calvinist missionary preached to a clutch of islanders. Sal was inspired by this skeletal figure, atrophied by disease while his faith grew flesh, preaching to a group of Hawaiians who didn't even understand what he said. He won a convert in Sal who regained his faith, albeit a protestant one, in that roofless church with the wind howling through it.

Faith regained elicits stronger emotion than getting hooked the first time. Sal was dewy-eyed as we walked in the house. His sons shuffled around, embarrassed by their father's tears. When Jacob could take no more, he said, 'That was the year Bell filed an application for a telephone patent,' underlining with something neutral his father's galling tale. He shook his arms out of his coat, his hair brushing the ceiling. 'Same year exactly.'

All that night the chains over the river rattled in the wind. I lay awake in my cot while Jacob and Paul were asleep, listening to the lambs crying, trapped in the hanging steel. Later still I heard Leah stumble across the darkened threshold and up to bed.

43

2 NOVEMBER 1959

There is a rawness in the air left behind after a storm. Vestiges of snow lie on the ground. The mountain paths are iced up after a downpour of rain. It takes near an hour to walk to the bluff, where the sight of the surf raises hairs on the back of my neck. At mid-tide the rock is exposed and waves burst over its craggy back. The shorebreak scours the sand, ripping out big stones from their sockets. Sea vapour hangs in the air.

Soaked by the outgoing tide the sand dries in the shape of the North American map. It's what I see, anyhow. Gorse growing on the cliff reflects in the sand and lends the Californian coast in the Pacific Northwest a green tint. The Deep South comes on black, stressed with white faults, sparry veins.

With my field-glasses I locate the surfer turning in the gyre of cross-currents. He is having the same battle in hell getting out as yesterday, until he leaves the treacherous shorebreak behind him. Safely through the outer sandbar he circles in front of the rock. The rock is covered in sharp mussel and whelk shells, like a wall crowned with broken glass. Ships have been wrecked on this rock, including a slave ship carrying slates for ballast. All hands went down. Children's bodies were washed up on the beach for days. Which is how that piece of rock earned its name ... Moloch.

He spends a long time shuttling about for position, fighting the westerly rip. He bottles out as the first set looms, and paddles to get out of range. A swell lifts him skywards and drenches him in the twenty-feet veil of spray it sends back in a full spectrum of light. Seconds later the wave takes a fairground ride over the rock and gives it such a pounding that bits of limestone fly off into the air.

Cormorants glide inches above the surface of the water. He is joined by a couple of humorous seals who keep popping up to see why he hasn't caught a wave. It has to be very spooky out there alone with the sea creatures and the spirits of shipwrecked sailors hissing from the black fissures. Thirty feet in front of him the rock looks vindictive, misty and crab infested, its kelp antennae reaching into each wave.

A wave closes in and he starts to run before it, then backs off at the last second. He lets that perfect wave pass and I comprehend how frightened he must be out there.

I catch a glimpse of his face through the glasses. He looks too young to fight this rough and I begin to fear for him. If he squandered the whole day out there waiting for the groundswell to shrink, I'd think no less of him.

I bet it's noisy out there. It's noisy enough up here, the explosion of waves louder than any quarry blasting. If there could be a slackening off of the noise, just that, then I guess he might go for one.

He should get angry with himself. A state of mind is what he's been hanging around for, all this time.

He paddles over the first wave in a huge set. Cuts through the feathering lip of the second to be greeted by an insecure structure fourteen feet in height. He has no choice but to ride the avalanche out of trouble.

All the danger is in the first few seconds, in the take-off. He stands and drops down into the trough, turns, splitting open the right shoulder. My shoulders relax the moment he clears the rock. The one limitation of Moloch is it can't get up and chase him.

I follow his game plan with interest. After completing all the opening moves he carves up the wave. With his arm stuck in the wall he glances up. A roof forms and encloses him. He reappears on the shoulder of the wave, shifts his left foot to the outside gunwale and cuts down the face, returning to where he started, in the turbine house, the nerve centre. I see that he is heading directly for the rock, about to lasso him in its kelp. A one-eighty-degree turn brings him round underneath a new section of wave. A second roof forms and he hurls through sizzling rain. He is in instinctual territory now, reacting to the speed of events, to the pile-up of water, to its schedule of collapsing power.

The ride lasts no more than ten seconds.

This is marvellous theatre. Somehow this boy has found a way down from off the mountain and claimed a bargain in the sea. For sure he has a quarryman's instincts. The big difference is, the mountain never moved this fast. The sea migrates and he travels with it, guest to its power, while the rest of the world here has stopped.

Surfing is primitive art. What else are waves for, that have been pumping into shore since the thaw of the ice age? Surf was put there for us to play with. Nature gives us the means if not the instructions. Think of the horse. The domestic horse for thousands of years was kept for its meat until someone mounted one.

It is an atavistic performance. I mean in particular the act of *standing* on a board. A princely act that links him with eighteenth-century Hawaiian kings, but even further back altogether to the origin of species, when man's ancestor first got to his feet and walked out of the sea.

A dozen more waves, a dozen guesses. He attempts on

each ride to reclaim that single moment of grace when he is inside the belly of the wave.

It is getting close to low tide now and the limestone shelf running alongside Moloch is exposed in places. The waves have started dumping down their entire length, the rides shorter and crazier. They have become impossible to negotiate. He paddles around the break and rides the white water across the sandbars and through the shorebreak.

I watch him walk up the beach to where a scattering of framed canvas huts left behind from summer are anchored against the cliffs. He selects the sturdiest of them and crawls under its flap.

I head down to the beach.

From here the waves are far bigger, noisier and look less negotiable than I originally thought. Perspective changes, nothing makes clear sense any longer and I get a lesson in insignificance. My excitement turns to moroseness. The ground trembles from the power of dumping shorebreak. It is loud all the time. Waves make so much noise, ripping like canvas, blanking out the rest of nature. The mountain on which I stood a while ago is a jumble of slopes and screes, gunsmoked gorges and cloud-filled plateaus connecting peaks, the green and spar and heather blue seams of slate perfectly mirrored by the sea. I study the land, the quarry, with sad longing. They seem as far away and unreachable as history.

At dusk the waves make beautiful forms. The tide has gone past its lowest ebb and is creeping back over the shelf. The wind has slackened leaving the sea black and smooth as polished glass. Waves appear out of the twilight, purple against a blood-red sky. From a long way down the beach I see the boy reappear from the tent, enticed back into the water.

47

The shorebreak is kinder at this hour. With the drop in wind the sea is less raw. It has gained a certain moderacy. Although less hollow, the evening waves are exhilarating in a different way, spectral and enigmatic in the twilight. He paddles out to the rock and this time doesn't wait around but introduces himself to the first set in. Watching him from the beach now, he keeps vanishing from my sight behind the swells.

After half an hour or so I achieve a state of deep calm. The blueprints of each of his waves, ridden and tamed, pass across my mind. When I see him again he has drifted nearer to Moloch. A wave is close to breaking on his back. He spins the board and strokes hard towards the wave. He passes through the lip. The second wave feathers dangerously. The wave behind that will be bigger still; he knows he has no choice but to surf it out of trouble. He turns the board and falls across the deck, stroking once, twice. His take-off is late. Tons of water follow him down the face where he turns fast in response to desperate conditions. He pulls too far round, in an arc of fire. The fin pops out and the board spins, returning him back up the face of the wave, to meet head-on the sheet of freefalling water. He loses it all in less than a second and drops over the falls, separated from his transport.

He suffers out of sight, out of my reach. The following seconds are unbearable ones. I fear for him as though he were my own son. The wave won't let him up; the turbulence of tons of collapsing water buffets him. There is nothing he can do at a moment like this except journey at its discretion. The only way to escape the violence above is to try to sink deeper into ever-darkening and colder water.

I glimpse him being bounced on to the rock. Tears

blossom in my eyes. This is the way those sailors went. I appeal to the inanimate rock with its kelp beard to let him up for air. But the Moloch refuses to hear my prayers. It has no flaws or human errors to exploit.

Through the glasses I see his hand trawl the long shaggy kelp, trying to get a hold. The wave makes one final grasp at his legs then churns inexorably on. The water drains off the rock, leaving him beached and flailing. The third wave of the set inches in, close enough for him to reach out and touch its face. He tightens his hold on the kelp. I take a deep breath with him.

The wave comes down with a long whistling howl. The kelp rips out of his hand and he ricochets off a peak in the rock before vanishing. I close my eyes and feel a tremendous tearing at my own limbs. My head is yanked backwards, oxygen is pounded out of me. My eyes still closed, we are canopied under the same dark sky. The same stars that are in his eyes are in mine.

By now the agitation will have slackened off to a gentler, swaying cycle. This is the moment to make his ascent. I stare through the glasses at an unbroken surface of water. His foot must have caught in the weed. He will have no resources left. He will start to feel the fatal relief of defeat. My helplessness feels like an illness, like a sin. I look around for assistance, but I am alone on the beach.

The water suddenly drains off the rock and I see him there. And his legs *are* chained to the kelp. He fights to release himself while the next wave approaches at twenty miles an hour, trembling in anticipation of its little human snack. He frees his feet just in time and dives off the rock into deep water.

His board is lost to him as he struggles in the smoke and unreason of the sea. He has nothing left to give, no power

49

in his arms or legs and drifts in on the waves. His mind will be wandering, as men's minds wander in a desert. Lying on his back, he scoops the water lazily. Time owns him for a little while. He doesn't fight the current but intelligently lets the shorebreak take him in, those rough babysitters. He reaches the beach less by his own effort, than by the indifferent grace of the sea.

Beached a hundred yards away, his board sits indignantly on its fin.

He limps up the beach to the tent and crawls in under the flap. The sea roars on in twilight.

I make my way across the sand. Outside the tent, stones make brittle percussive notes underfoot. The tent platform heaves as I climb on.

'Who is that?' His voice sounds young and frail and cannot conceal his fear. The canvas shakes gently as I force the flap. The air inside grows stiff. I can just make out the shapes of deck-chairs, but not of him.

My voice shuttles through the dark. 'Mind if I join you? It's freezing out here.'

'It's not my tent,' he tries feebly.

I go in anyway, crawling under the flap, scraping my belly against the floorboards. I still can't see him, but I can smell him. No sooner am I in before he reaches out and touches me. It's been a day for acting on instincts. He seems reassured that I'm not made of more limestone. His hand traces its way slowly up my chest, over the buttons of my jacket, around the collar of my shirt. He moves closer, brushes his arm against mine. I feel hair in my face. Our anticipatory breathing fills the tent. Already exhilarated from survival he wants everything else that is natural to happen. He has earned the right. He is alive when he should have been dead. I force my breath to play in tune with his and we meet in the space above damp towels. Darkness cloaks inhibition. I peel his jacket off, made of nylon. There is a thin layer of rubber inside. Then I roll down his trousers, also lined with rubber. I fill my cheeks with him and he follows my lead, enclosing me with his mouth. We form an unbroken circuit, a perfect symmetry. He smells of salt, seaweed.

We accelerate and retard each other's progress down the line. He swells and bucks inside my mouth. Adam upon Adam, rubbing out half the world. His seed puts enormous pressure against the back of my throat. I send his own fluid back in a loop. Our barks trace a line across the sound of surf outside.

Two strangers realizing one other; it's a found poem. It is God's country to know exactly what two strangers need without asking. 'Aaron Lewis,' I say in the cooling. 'How do you do.'

'Mine's Glanmor. Pleased to meet you.'

There is a wicker basket in the tent that is filled with tins. He fumbles around for an opener and breaks into three of the cans. We cleanse our mouths with peaches, baked beans and prunes. 'You're a foreigner,' he says.

'Got a problem with that?'

'No.'

'Must have felt scary out there in the surf.'

'I didn't feel anything, one way or the other.'

'It looked frightening from where I was standing.'

'Emotion has nothing to do with it.'

'You licked it though, didn't you? You won't be scared of me then?'

'I'm not scared,' he says, trying to sound tougher than he is. 'How old are you?'

'Old enough to be your father probably. How old are you?'

'Seventeen. Where do you come from?'

'The United States.'

'You don't sound American. I thought you were German.'

'No, I'm American.'

'One day I'm going to go to California. Huntington beach. Malibu.'

'You know your geography.'

'I like geography. It's history I don't like.'

'Yeah, how come?'

'History is cruel to people. That's why I like the sea. The sea is a wilderness.'

'And there is no history in a wilderness.'

'The past is the wave I've just ridden. The future the next wave coming in.'

'There is something very familiar about you. What's your last name?'

'Gravano.'

'Uh huh. Any relation to Paul Gravano?'

'That's my father.'

I gave myself a moment to recover. 'He and I go way back. So who's your mother?'

'She died.'

'I'm sorry.'

'She died having me. I killed her. He hates me for killing her. He won't admit it, but he does. Even gave me a secular name. When I was drowning out there a moment ago I saw her. She was just getting home from a dance, wearing a fox stole. I watched from the top of the stairs. My father helped her off with her coat. And she said, "I wonder if Glanmor is sleeping?" I heard her speak my name.'

'My father was a regular kind of guy. My mother out-shined him. How two such different people end up pitching camp together, I just don't know. When he died I thought what a shame. When she died I was totalled.'

'How do you know my father? Did you work in the quarry?'

'You bet I did. I even lived with him for a while. You still up in Red Lion Road?'

'Yes.'

'I want to see your father. I want to tell him so many things.'

'What things?'

'I don't know, everything, I guess. I just don't have the words to fit the emotion right now.'

'I'm surprised to see this stream still running away past the house,' I tell him. In a minute I'm going to meet Paul and I'm bereft of a plan of action. What do I say? Who will speak first? Why do I feel such a frail sister? I used to have at least three openings filed away for the unexpected. I look again to the river bubbling in the dark, about the only thing in Sharon that hasn't dried up. 'As a lodger here I used to watch the rats out of the bedroom window climb off the banks with pieces of sheep's liver in their teeth. We pissed in this river and smallholders operating their own abattoirs dumped offal into it. It was a serpent of disease. Drink from it and you'd die as sure as if a rattlesnake had bitten you in the throat.'

Outside the house a few scrawny chickens loiter in a slate enclosure and piled up behind the back door are bundles of firewood. Glanmor tells me Paul sells the firewood. 'Since that strike he's never worked again in a proper job. He just hobbles around town selling firewood, raising chickens, raising me. But not as well as those fucking chickens.'

The house is full of smoke as we enter. But there is no leak in the fire, all the smoke comes out of me. My hands are trembling so much I hide them away in my pockets. I can't get past the doorway, there is an embarrassing breakdown in motion. Then I see Paul sitting in a chair by the stove, listening to music on the radio, the 'wireless'. My mouth drops open. He has hardly changed at all. His skin is taut and shaved. If anything, without his beard, he looks younger than when I last saw him. He turns his head towards us as Glanmor pushes me into the room.

His eyes are as cold as the stone in my mother's sapphire ring. What is wrong here? Why doesn't he welcome me? He was never short of something to say, as I recall.

'Father, I've brought an old friend to see you.'

'Who is that?'

'It's me,' I say laughing. For sure he's making a joke. 'Have I changed that much?'

'Aaron Lewis, is that you?'

'Paul . . .'

'I recognize your voice . . .'

'My *voice*? You haven't aged a day. How come you've survived so well?'

'The faith has kept me young.'

I notice a shine in his eye, something glinting. He is suspended in a curious mood, stroking his chin, refusing to say any more. He has not seen me in twenty years and all he can say is, I recognize your voice. I walk deeper down the line until I am at his side. I touch him on the shoulder. His flesh sends back an alien message. Then I realize. He can't see me. Paul is blind.

I turn on Glanmor angrily. Why didn't he warn me? 'When did you lose your sight?' I ask in a passive voice.

'He lost it gradually,' Glanmor answers for him. 'Cataracts.'

'Can you see anything?'

'Nothing. But at least I remember what you look like, Aaron. People I've met since . . . All I have to go on are their names.'

'Paul, I'm sorry. *Damn!*'

'It's all right. Being blind isn't so bad. It's private.'

His books are piled up on their sides to the ceiling rafters, a lot more than I remember. Books salvaged from abandoned quarrymen's homes, the ballast that impeded flight. Books were so much a part of his world. Now they lay redundant around him. In a corner of the room is a bed, heavily dented in the middle and covered in a light blue patchwork quilt. It looks as if Paul has been camping out in the parlour.

'What have you been doing with yourself all these years, Aaron?'

'Going to bed early.'

Paul takes a log off a pile and feeds it through the gate on the top of the stove. The fire greedily consumes the wood as I browse through some of the book titles: David Brewster's *Letters on Natural Magic*; *The Life of William Robertson Smith*; *R. C. Morgan: His Life and Times*; *The Life and Personal Recollections of Samuel Garratt*; *The Preacher and the Modern Mind* by George Jackson; *The Persecution of George Jackson.*

The silence threatens to settle in all night, as if nothing will penetrate it. This is what long absence does to friends: it destroys their language. Glanmor tries to cover up by making tea. He heaps three spoonfuls into a ceramic pot

55

heated from the range, and pours in boiling water from the kettle that simmers on the stove. We wait for the long minutes to pass while it draws, then for Glanmor to pour out three cups. Paul removes his hand from his lap under the table and feels around the curve of the cup. He tests it for heat before lifting it in both hands to his mouth. He replaces the cup not quite squarely on the saucer and then his right hand goes searching for the cake that has not yet materialized. Glanmor notices this, that his father is one step ahead, and quickly fills the space in front of him with fruit cake. He nudges his father's wandering fingers with the edge of the plate. Paul touches the cake, caresses it all over, measuring size. It is quite a lump that he senses and he breaks off a piece. He opens his mouth and carefully steers the cake in, the crumbs falling into his lap. Like a horse he champs down for a long time before swallowing, then goes in search of the plate again.

'Don't you have a job to go to?' Paul asks his son.

'It's six o'clock at night.'

Paul changes at random the wavelength on the radio until he finds a station giving the news. A crisp and impeccable English voice puffs up the room with strife. Gangs of unemployed youths are rioting in Cardiff, Liverpool, London – burning shops, cars, houses, their own youth clubs. He makes an aggressive gesture with his hand, snapping the radio off. 'The youth of today act as if the Ten Commandments are negotiable,' Paul retorts.

'They're just poor black kids that are rioting,' Glanmor says. 'They don't go in for all this Union Jack waving stuff.'

'It's an insult to the poor who have raised families with dignity to say that all you can expect from poor people is that their children will become hooligans and thieves.'

It's comforting hearing him talk politics. Everything is

all right, everything is normal for the moment. Paul's skin reddens, the muscles in his neck start straining as he stretches into a folksy pose. I recognize all the warning signs before Glanmor does. He is about to deliver. 'Young people today award themselves freedom to do as their conscience tells them.'

'Isn't that a good thing?'

'Let me tell you something, Glanmor. Once upon a time in this town there was so little for a police officer to do he had to go touting for business. When some destitute Irishmen landed in Sharon he followed them over the mountain ridge, pushing his bicycle all the way. He waited until they got to Bethel before arresting them for vagrancy. There were no police cells in Sharon. There was never any call for them. There was natural law in operation.'

I'm getting a strong impression of the trouble here. He likes to torment his only son with politics. Except politics is abstract now, no longer a real-life issue as when so much of the conflict lay just fifteen hundred feet up in the mountains. Strong local politics has gone by the way.

But Glanmor isn't interested in politics any more than he is in history. It is his body that has all the best stories. Body politic. His generation has just discovered that the body has more than one orifice. Paul calls this newfound freedom pseudo-freedom. 'All modern children have a dash of vice in their blood. They have relinquished faith, and weakness is their talent.'

A little pleasure of that kind might have eased Paul's burden, softened him up after all those years spent in the cold air. That silent frozen space around the quarry has made him severe. Paul has become a blind tyrant with the one browbeaten subject. But, despite the abuse he piles upon him, Glanmor is all he's got since his sister Leah fled the nest and his brother Jacob perished in obscurity in the coal fields of Schuylkill County, Penn.

Glanmor and this house is all he has left. It is his ark and the flood never more than a few days off.

Glanmor clears the table of his father's empty plate and begins to scrub the surface as I remember Rebekah doing, in long even strokes. It is the same table that has been in the house for over seventy years, scrubbed daily by two generations of women and now by Glanmor. Its soft maple surface is uneven and grooved from the action of the bristles. The woodstain has long been bleached off, its legs weak and wobbly. People came to the house and told stories, each tale falling upon the scrubbed surface under the weight of its own gravity. If I were an artist I'd paint the subjects of those stories and locate them *on* a table top.

He goes down on his knees to wash the slate doorstep, cleaning away one day's shoeprints in order to make way for the next's. Again, he looks the embodiment of his grand-mother. Inventors of the Formica top, vacuum cleaners, have failed to take into account the sacramental nature of such chores in their bid to free women from domesticity. If you can't keep your house clean when all about you is falling apart, then there is nothing.

Paul and I shoot the breeze, in particular about how trade unions have burgeoned since the war. 'Unions have the dimensions of armies now,' he says. 'And in every one of those men there is an Adam.'

Can it be possible that, while I have shifted my once-fixed beliefs, Paul has moved in to occupy my old ground?

'Unions today demand higher wages as a recognition of their worth. But when productivity drops, by the same argument, it is right they should voluntarily lower their price. But they never relinquish their gains. Greed has replaced pride. It's a course set for self-destruction. One day the clock will be turned back on the unions.'

'Well, that was very fucking strange,' I tell Glanmor, who is walking me back to my guest-house. 'You could have warned me he was blind.'

'He's getting old.'

'He's the same age as me.'

'You don't look it.'

'Oh, I bet you say that to all the girls.'

Outside the guest-house Glanmor's lips part in anticipation of a kiss. I shake his hand chastely, so as not to get tongues wagging. He has to live here after I've gone. Such coyness seems to worry him. 'How long are you staying?' he asks. 'Can I see you again?'

'Come in now if you like. I'll give you a nightcap.'

Inside my room Glanmor looks at his reflection in the dresser mirror and asks whether I ever met his mother. 'What was her maiden name?' I ask.

'Juliet *Bowne*, I think.'

'You think?' I say, remembering her well. She was just a kid then, a leader of children. She led them out of childhood into adulthood. I studied her form in the cemetery one time, drinking her father's moonshine with her friends, yanking the flagon out of each other's hands in a parody of drunks. The drinking done, she instigated a kissing competition. Picking one of the boys a whole foot shorter than herself she leant him against a headstone and almost asphyxiated him, while the other kids shouted at the top of their voices: 'Go on Juliet! Go on Moses! You got to beat three minutes.' She could have been one of Leah's protégées, she had the same depth of appetite. But how the psychology worked, regarding Paul marrying his sister, I just don't know. 'I don't think I ever knew her. But tell me about Paul. How has he managed to nurture you without a woman's help?'

'He did a better job of it on the mountains than he did in the home, I'll tell you that for nothing. Dad raised me to be a walker.'

He tells me how they always set off at the same time just after dawn with a pack of sandwiches and a flask of tea. The quarry was as far as Glanmor ever wanted to go but Paul preferred not to linger there and set the pace, ignoring his son's plea to stop all the time, knowing that he would eventually get beyond his initial fatigue and establish a rhythm maintainable all day. His eyesight was going even then. If he was fanatical about walking, it might have had something to do with his fear of never being able to do it again.

They would walk for hours before stopping at three thousand feet. Paul would get him to look back at Sharon and gauge the massive distance made, query the insignificance of human endeavour. 'He'd tell me to put my hand out so I could fit our town into my palm.' It was easier to love his father out there. As long as Paul was walking he seemed happy and his happiness rubbed off. He was free of polemic on the move, with a rucksack on his back, a staff trembling under his weight. All the bad history poured out from the backs of his heels.

A regular haunt to stop and eat lunch was a Roman legionnaire's lookout, now an eagle's hunting ground, where the silence was as tangible as a sheet of glass. Stationary, Paul would start sounding off again and spoil the perfect stillness as Glanmor devoured the sandwiches. 'If I'm a fast eater now, it's because of that.'

We have our nightcap. Afterwards, drawn and loaded

and sleek with sweat, Glanmor sweetly rests his head on my chest. Outside the window the mountain range is translucent blue. The quarry shimmers in starlight. A bird of prey falls out of the sky and plunges into the coliseum of slate.

21/12/37

Dear Sir,
I have a small lead. Yesterday at dusk I watched two rockmen in my team smear a section of rock with a clear oil. One man held a candle a foot from the surface while the other man moved his sightlines, observing the reflection of the candle in the oil. They spent a great deal of time doing this, moving right along the surface both horizontally and vertically. I made inquiries of them as to what they might be doing and they jested with me. 'Looking for ghosts', is what they said. I suspect that what it was I saw is of some significance.
Respectfully submitted,
A. L.

WINTER 1938

At 0600 on the fifteenth of February all eighteen teams working on Sevastopol gallery were ordered to report for work to a contractor. The contractor was to be in sole charge of the gallery, employing men as he saw fit, paying for their wages, gunpowder, horse and haulage expenses from his commission. I had no idea it was happening until it was happening. Nor anyone else. The awareness that a two-hundred-year-old practice was going quickly spread around the rest of quarry. Crews from the Red Lion and New York galleries flooded Sevastopol. The air was filled with a misty, sepia rain sweeping diagonally. The rain softened the vistas, but not the emotion, that singular thrust for survival.

'Contractor? What is contractor?' Sal was trying to read the notice signed by Manning that we'd torn down off the quarry gates. Bodies stampeded past us and knocked us every which way.

'Manning's replacing the bargain with a prototype contract system,' Paul said before sheering off into the crowd. The bargain was something they had devised for themselves. It was at the heart of their culture.

His face burning red, Jacob marched off to where the contractor was sitting calmly inside the penthouse on the Fitzroy level, mistaking the mass approach as a submissive clamouring for work. Through the window we could see him huddled over quarry charts, he and his two sons looking infernal in the candlelight. Jacob entered the shed and a moment later the contractor was ejected from the doorway, like a fox rustled from its den. Ishmael, the blind smithy, restrained the sons in the doorway as Jacob spun the contractor to the ground.

'I am asking you levelly. Will you do the correct thing and leave the quarry?'

The contractor twisted his neck around to face Jacob, but did not answer the question. Jacob brought him to his feet and punched him summarily in the face, sending him sprawling again. He lay on the ground, inanimate as a branch from a tree brought crashing down by lightning; coiled tightly and without soul. Paul stamped out of the crowd and seized his brother's arm. 'This is wrong. This is very wrong,' he shouted. 'We will be punished for this.'

Jacob's face was closed for business. I'd seen that kind of look on men whose children are taken away from them by their ex-wives; on horse trainers who find their thorough-breds skewered through the heart by saboteurs. It is a rage so strong that it consumes the blood and has no reason.

I pulled Paul away from him. 'You take care of the diplomacy later,' I said.

Some of the men started kicking the contractor as he lay on the ground. His sons broke free from Ishmael and fell across their father, protecting him with their own bodies, an impressive act of love in the form of tangled flesh. They never murmured once as boots connected with their ribs, thighs, heads. Paul shouted until he lost his voice.

It was not for me to make moral judgements over any of this. I was keeping neutral, staying right out of it. While broadly sympathetic with Paul's view, I was also keen to see what was left of the contractor after they let him up.

It could have been me down there so very easily.

The contractor was paraded through Sevastopol, New

York, Red Lion, Springtime, Blue Lion, Jolly Boy galleries, his legs buckling under him, leaving splashes of blood on slate tableaus. Wild cowardly punches were thrown. Men waved their bowler hats from across the lake. Eventually he was escorted off the mountain and delivered, limp as a freshly killed buck, into the arms of his wife.

I spotted Edward Manning trying to check out early. He was making a dash for the river, disguising a run as a brisk walk which threw his shoulders out, black velvet cloak luffing in his slipstream. My mouth went dry watching his bid for safety.

I didn't think my next reaction through, it was all in my legs. I ran after him, pursued him outside the quarry along the river bank, dodging ghostly silver birches and leaping exposed roots. I brought him down with a flying tackle, my arms around his neck and both of us tumbled into the river. We made a hard and bitter landing in icy water. He struggled to free himself, throwing wild punches that connected vaguely with my back. Holding him under the water I watched the casual way his velvet cape floated up to the surface, watched his scheduled death take shape in his face. I wanted him totalled. It was just my way of prolonging the pleasure, was all. I let him have a sip of air then pushed him under again.

Paul was shouting from somewhere on the river bank, like the voice of the lawman out on the porch, heard through your sleep. His voice grew shriller until I diffidently twisted my neck and sought out his face. He was kneeling on the earth, his flaring eyes burning holes in the monochrome landscape. My motivation started slipping. I pulled Manning out of the water and beached him on the bank. He was snatched away by several rough and outstretched arms and dragged along on his knees through damp moss. I was left holding Manning's waterlogged cape in my hands, feeling bereft. The river tugged at my legs like a demanding child.

Neither the contractor nor Edward Manning returned to the quarry the following day, Friday. The quarry was skipperless and Jacob had reason to believe his violence *had* brought in results. We worked all next day in fine weather, with no twinge of fear of retribution, divine or other, on its way up to greet us.

That Sunday Mr Parry's sermon was on apostasy. After service he celebrated with the men in the streets. Then on the Monday Jacob was issued with a summons to appear before the magistrates court in Port True, charged with assaulting the contractor.

'Is there nothing for me in that postbag?' I asked.

'Perhaps Manning could not see your face through the rushing stream,' Jacob jested.

'Maybe he thought I was baptizing him.'

'You have a bad seed in you, Aaron,' Paul said. 'I'm going to have to keep a close eye on you from now on, lest it get out of hand.'

The quarrymen's committee met that evening in Sal's house. Saul P. Howells advised Jacob to submit himself to the authorities without resistance. 'Eleven warrants have been issued in total, Jacob. We've told everyone to go. A Manchester lawyer will be defending you in court.'

'A lawyer has to be expensive,' Rebekah said from the head of the table, Leah's baby squirming on her lap. 'Who will pay his fee?' She was wearing her usual bright colours, yellow and orange, that embarrassed the dark-suited fraternity. They shifted around uncomfortably, not used to having to answer a woman's questions.

'That's taken care of.'

'By whom?' she asked.

'The union.'

'The union is a grand expression for a few subscriptions, Saul. How can it afford to pay a big lawyer like this?' Rebekah kept up the heat.

For a moment it looked as if Sal was about to ask her to leave. But he backed down when she got up from the table to walk the baby around the room. She rocked the infant so energetically the kid was frightened into silence. Her loose grey hair swerved around her face like a cart taking a corner too fast.

Rebekah always moved about her own house with a hard-earned authority, a clay pipe in her mouth and some kitchen utensil in her hand. Her energy sometimes verged on the violent, smashing into furniture as she went. She once said that her greatest fear was losing that energy. 'One thing a woman has to learn fast in this place. These men don't do a thing around the house. Lucky if you get them to lift a fork to their mouths. If I lost my energy I'd be up to my neck in dirt in no time. Leah helps out but has to be pushed into it.' She gave me a soft smile. 'You've settled in nicely here. I could tell about you when you came. Our house is small but it isn't claustrophobic. We live on top of one another but you'll have found that your thoughts are your own. Your thoughts are the only space you'll be getting around here. Guaranteed.'

The magistrates court was situated fourteen clicks due north, in the port town from where Elusen shipped his slates to destinations all over the world. The eleven men left Sharon at four in the morning and covered the distance on foot, accompanied by an escort of five hundred Sharonites.

The procession had a carnival atmosphere with a silver band playing at the head, the men walking with swing. For those fourteen miles I bathed in their optimism, under a warm crisp sky, group movement lending me an omniscience. There's a whole lot to be said about losing yourself like that, in a big angry crowd. Bad men and good take on the same generic character on the march, creating an electricity of a kind; hobnail boots scraping against the road. I can't say I've felt happier or stronger. What I lacked in personal morality was made up for me in numerical strength. But the strength lasts only as long as you keep moving.

The magistrates acquitted seven of the eleven men. Jacob, Ishmael and twin brothers were fined between four and six pounds for causing grievous bodily harm. The hearing lasted three hours; the money for their fines raised in three minutes outside on the courthouse steps.

All eleven men, including the seven acquitted, were denied entry into the quarry the next morning by four new and hard-faced agents mounted on grey geldings. As I wondered where this whole episode would lead, I spotted Edward Manning on foot, his rosy-cheeked womanly face bobbing between horses' withers. He looked a lot healthier than when I last saw him.

Trembling slightly, one of the twins grabbed a horse bridle and tried to appeal to Manning. 'I have worked in this quarry for forty years. This is the first time I've been in conflict with any official.' When Manning failed to respond he added, 'I cannot face my debts without work.'

'Then walk backwards to meet them,' Manning replied.

The road behind the agents was filling with an army of moleskinned quarrymen who had just discovered contractors were in place on every gallery. The agents looked nervous and pulled their horses' heads round and galloped into the quarry, parting the crowds of men staging their walk-out.

A strike ballot was held outside Jerusalem chapel. Votes counted were 2,170 for, 77 against and 800 abstaining. A great fire was made of the ballot papers to ensure the names didn't get into the wrong hands. The minister of Jerusalem put the torch to the bonfire. I stood next to him watching the flames devour 2,170 contentious names and rise into the sky as ash.

We marched through the town and into the mountains in one unbroken column. Once again, I shared their singular resolution and felt the power return, a luxurious confidence, hitherto unfelt. We walked on in the rain, collecting ever more people into our ranks at satellite towns of Nebo, Carmel, Bethel. Those too old to walk cheered us from their front porches. Down into the pass the shoulders of the mountains sheered up protectively. Close to our skin the land held out its yield: purple thrift, early purple orchid, heather, foxgloves, bloody cranesbill, viper's bugloss, golden bracken – like bouquets proffered to returning armies. Goats stood sentinel on loose rocks. Streams of white water fell down the mountainside. Dogs working flocks into lower pastures moved around our heads. Rain flooded the valley and anointed us. Spiralling hawks vanished into the mist. The mist was buoyed between layered escarpments. Escarpments were echoes.

The marching stopped and I became painfully aware of my isolation. For days following the walk-out I was chronically homesick. Bent over double, to be precise, with colic ... melancholia. To distract myself I fantasized about going to the track. A horse race always cheered me up. That too was a mass in motion. An outing to Saratoga Springs would have been welcome, to be surrounded by gamblers for a change. Saratoga was a classy track, but I'd have settled for a local meet had there been one. But there was no track for miles. You couldn't even get the odds in Sharon for races as far away as Newbury. The newsagent, a pious evangelical with twelve children, inked out the form in every newspaper before putting them on the stand in the morning.

I walked my despondency into a state of anger through the woods. By the time I reached Lord Elusen's house, overlooking the Menai Strait, an hour later I was ready to fight anyone.

The ceiling to the entrance gallery was slung so low as to subdue the visitor, depress his spirits a few moments so as to increase the impact of stepping into the hall. That hall was really something. Light seeping through stained-glass windows emblazoned the thirty-foot walls. The rugs were so vivacious I kept seeing leaf patterns swim in front of my eyes. A fire burning in the distance looked like a candle. That's how big the place was. I surprised a butler standing on a ladder, dusting a huge gilt-edged mirror above the fireplace, the squeaking of his rag echoing in the ceiling bosses. He stepped down from the ladder and approached me under a fast pressure gait, fiercely checking out my dress, prejudging me. Before he got within range I let him know, in a borrowed South Carolina accent, that I had come to see the boss. Reluctantly he showed me into a drawing-room off the hall.

The door was closed behind me and a stillness fell over the contents of the room like a soft dust. I felt calmer than the moment before. The room was decorated in yellows and pinks with silk hangings and curtains. A central candelabra spilled an uncertain light over everything.

The butler returned and led me into the library. I saw initially a young girl playing a harpsichord, then a man, stretched full tilt across the billiard table, who turned out to be Elusen himself. The bed, frame, pockets and legs of the billiard table were all made of enamelled slate. Elusen eased his cue back and forth through the bridge of his fingers, lining up a transit with the red ball and centre pocket with a solemnity and seriousness of purpose as though nothing he would do for the rest of that day could be as important. In a grey suit, he was as intense and lithe a figure as his portraits hanging around the house suggested.

The butler held his hand parallel to the floor, indicating that I must wait for the master to execute his shot before there was any possibility of an introduction. Stone faces stared at me around blind sides of limestone pillars, gruesome masks bearing fierce teeth, each one a predator, killing or being killed. Carved into the legs of a desk, eagles rode bulls in a fight to the death. In glass domes on window shelves was a stuffed Indian roller, an African bush shrike and a mangrove kingfisher. No respite anywhere from the images of prey. I began craving a plain room with an object or two of pleasure: a sofa and a liquor cabinet.

He fired the white on to a red and pocketed something. Since he now seemed to have a break going, this visitor decided to take a seat. I leant one arm against that desk supported by serpents and wild beasts. The butler glanced at me as though I was born out of wedlock, which may well be true, but even bastards need a chair. To hell with him. I wouldn't call his lordship's manners particularly exemplary.

Several opened letters lay on the desk. A carved ivory paper-knife lay across them. I moved the paper-knife to the side so I could read one of the letters.

Dear Father,
Thank you for my pen. It is truly superb. I love to write in black ink through 24-carat gold. Thank you also for the watch you kindly sent me. Its time-keeping is without reproach.
Your devoted son . . .

The break came to an end and I stood up from my chair. The butler tried to introduce me. Elusen responded by thrusting his cue in my hand. 'Play?'

'Not that kind of game, sir.'

Elusen registered more disappointment than the occasion merited. 'No one else in this house plays billiards but me and my son. And he is away at school. As you no doubt gathered from reading his letter.'

'I'm sorry, sir. It's a habit I've got into over the years.'

Elusen addressed his butler. 'Martins, fetch my guest some refreshment. Tea? Or something else?'

'Something else sounds fine. Whisky . . .'

We watched Martins leave the library. Elusen drew attention to his teenage daughter stabbing at the harpsichord keys. 'Some men claim that to be a quarryman you need musical gifts, Mr Lewis. By the sound of my daughter's efforts I would say it takes a lot longer to learn to play a harpsichord well. Jennifer, would you please stop.' His daughter's hands froze in mid-air. 'Better,' said Elusen. 'Felt like a damn church.'

His daughter glided away in the opposite direction, skirts muffling her footwork. 'Fine house,' I said. 'Must take a lot of upkeep, though. Whoever cleans your windows has a full-time job.'

'It's built in the style of a thirteenth-century fortress. In the time of Plantagenet Edward the First. It's meant to be siege-proof, but I understand you got in without any trouble.'

He leant his billiard cue against the fingers of a cast-iron lamp bracket shaped into a human hand and pushed the remaining ball into the corner pocket. 'I'm surprised to see you here, Mr Lewis.'

'Why didn't you let me know you were hiring a contractor?'

'There was nothing unusual in that. The men have always had a contract system. The bargain is a contract system. Except this one I control.'

'That's exactly why they rejected it.'

'We'll see about that.'

I joined him at the bay window and shared the panorama of a cool five thousand acres dipping into the Strait. I wondered how much the agency was soaking him for. Now I'd seen how this sybarite lived, I would reckon on, as a ballpark figure, a thousand bucks a month. I wouldn't open a file for less than that.

The house had no view of the quarry that I could see. Looking out over his land with his back to me, he said, 'Henry Pinkerton told me you were a man who knew his way around?' He turned and faced me, looking for a response. I gave him no satisfaction so he asked another. 'How is it going so far, Mr Lewis? I got your note about their optical experiments. Very curious, I thought.'

'That line of inquiry's been shut. The strike's put it on hold.'

Martins came through into the library with my whisky on a silver tray. I drained the lead crystal glass in a single mouthful. It was bitter stuff going down, like bourbon

73

coming up. Elusen watched me like one of his stuffed hawks, like my mother used to watch me eat. He began to resemble the granite bust he'd collected of the Roman emperor Trajan, who too built towns with his wealth. He followed my hand returning the glass to the tray before announcing that we were to take a stroll in the garden.

Outside, Elusen walked a couple of paces ahead of me. Pointing to a big greenhouse nestling between oak trees, he said, 'We grow nectarines, figs, peaches, strawberries, sweet peas, etcetera. All prize-winning stuff. When I'm at my house in London I have my fruit and vegetables sent up. Temperamentally are you a hunter, Mr Lewis, or a farmer?'

'Hunter, I guess.'

'We raise five hundred brace of grouse here in August. Three thousand pheasant between October and December.'

'That's a lot of pheasant. You eat all that?'

'Oh no!' he laughed. A frigidly composed walled garden came into view. Pruned rose-bushes, geraniums and fir trees corsetted in square plots. We wandered through as he continued his list. 'Last year my gamekeeper shot five hundred ravens, twenty peregrine and two hundred kestrels, protecting the game birds.'

'That's what's civilized about the English. The way they treat their animals like children. Before they eat them.'

He flicked on a smile like a switch. It went off again immediately. We walked by the stables. There were various saddle-horses and Arabs penned up inside, their beautiful heads peering out of their paddocks. 'Do you like horses, Mr Lewis?'

'Racehorses. I like racehorses. I was on the track a year before coming over.'

For fourteen months, as a matter of fact, tracking down the filly, McGuinness, that was rung in as Heretic at Morris Park and as Hiram Hines in New Orleans and again as Buck Wayne in Saratoga. In Saratoga I watched her come in on the nose at 40–1. I followed the guy I suspected was the ringer halfway around America – a lush in a three-piece Harris tweed, breast watch and cravat. When I saw him shooting in the paddock with a couple of known track plungers and a guy who pushes the queer in St Louis, then I knew for sure. I got it confirmed by tracing McGuinness's number tattooed on her upper inside lip back to when she passed through Missouri. Ringing was illegal in Missouri. They recorded each horse's tattoo and took a print of its nighteyes to go with the tattoo. The lush had bought McGuinness out of a selling race the year before and dyed her up the same as some old donkey.

'This might surprise you, Mr Lewis, but I hate this landscape. There are no hounds here. I think if there was fox-hunting it might have brought labourers and landlords together, on better terms. Let me explain something to you. Before my family came here, the locals put their lips to the earth and sucked out the moisture. They ate turf like animals.' His voice increased in volume, his complexion reddened. 'Two hundred years ago the people of these mountains were so desolate, etcetera, their existence was dominated by the need to find their daily bread.'

'They eat well now from what I see.'

A smile returned to his face. 'Thanks to my family. Have you had chance to notice the hospital?'

'It makes its presence felt.'

'My father built that. He built their homes too. They thought he was a fool.'

'I don't think they feel that about you, sir.'

'If I pulled my investment out of this country, they'd be left trying to turn bog into wine again.' He took several deep breaths. 'I tell you all this, Mr Lewis, so you may understand why I feel so abused!'

We walked back into the house, where he treated me to a whole other range of quarrymen's iniquities: the long-established practice of bribing his agents and foremen; the days of holiday they took collectively for harvests and funerals; their attempts to develop the quarry committee into a full union. For so long he had put up with all this on account of the way they could decide unilaterally the quality of the slate. 'I have never had any choice, just as my father and grandfather had no choice. Because they keep their skills secret. Because they conduct their business in a language I cannot understand. To my ears it sounds like a flock of pigeons taking flight.'

'I speak their language. It ain't so bad.'

'It took a long time for me to find a man with your qualifications. Your folk were local people?'

'Yessir. My grandfather worked in the slate mines near here. He emigrated when my father was nine. They ran the coal mines around Scranton a little in the same way. Same kind of secret codes. Different tune, same song.'

'You have no loyalty to them, nothing residual?'

'My loyalty is to the agency. The agency's belief is that employers have a right to know what their employees are saying and doing.'

'Is your father still in the coal mines?'

'No, sir. He had what most of these guys don't have: ambition. By the time I was born he was manager of the West Side Bank.'

'Ah, I see,' he said, but I don't think he understood for a

76

moment. He couldn't conceive of any man moving worlds like that.

'I know the nature of the men you employ, sir. I am of their stock.'

'Then you will know they are an idle people if given the chance. Like the African . . . Charming but lazy. I will not be undone in my quarry, Mr Lewis, I warn you.'

'It's a shame you can't change the ethnic mix here. It worked in Scranton.'

'Like a blood transfusion . . . Tell me, this strike. How long will it go on for, do you think?'

'It depends on what you want to do.'

'How long are you prepared to stay on?'

I thought steadily about what answer to give. I was within my rights to quit then. My original terms of employment, to reveal the quarrymen's art, had been thwarted. But there was something else that persisted in keeping me there, a power that wasn't political, which, try as I might, I couldn't override. 'Well, James McParlan was buried for over two years on the Maguire case,' I said.

His voice trembled at the scale of the problem. 'I am being held to ransom by my own employees and their self-appointed gang of little Bethelite preachers. It is a world I have lost control of, Mr Lewis. I am quite miserable with it all.'

'That's why you did good hiring private talent, sir. It's my job to restore the moral order.'

Nine hundred men and some of their families left Sharon in the first month of the strike, seeking work on farms and the coal industry in the south. Pickets stood at the quarry gates each day to antagonize the secessionists, the *cynfon* as they

77

were known locally. Elusen, on my advice, drafted in police to help the local cop manage the situation.

I went around to every public place and filled my ears. But it was in the Gravanos' house, where the strike committee met every week, that I learnt most. The Gravanos were my friends and hosts. I took regular checks on my conscience and found it in good health. There was no stone in my heart. The way I looked at it, the men had been content to take Elusen's money and hate him. He was my employer too, on this operation. I didn't hate him or anyone else, including myself. You start doing that in my business and you're up to your neck in shit in no time at all. That was the simple truth and I preferred to keep things simple. For this was not my fight. It was not for me to make moral decisions. I was not a priest.

Industrial unrest created an ecumenical climate in Sharon. General strike meetings were hosted by chapels in rotation and well over a thousand attended in the early stages. Ministers steeped their political sermons in Old Testament imagery. Before the strike it had been the other way round. Popular among them was the image of the tribes of Israel in the wilderness.

Into one such gathering at the Presbyterian chapel came an unexpected visitor in the shape of Edward Manning who had ridden into town on one of Elusen's Arabs. He staked the horse outside and executed, with self-conscious drama, his entry into chapel. Smoking a thin cigar and sweeping his new cape across his shoulders he walked down to the pulpit. People began shouting for his removal, calling him 'Elusen's grocery clerk' and 'brador'. He spun round under the pulpit, his cape sending a cold breeze over the front row. I shrunk some in my seat in the event that he'd come to finger the one who'd hounded him into the river. I respected him more for walking into the lions' den like that. He was earning a little character finally.

Manning adopted a defensive stance, pressing his back against a stone pillar. He eliminated the minister and the strike committee sitting in the deacons' chairs with his own sense of urgency. He read out a brief statement to the quarrymen: 'I, Lord Elusen, steadfastly refuse to meet delegates from the union or any such committee. I promise amnesty to any man who returns to work on the employer's terms.'

The laughter which followed him out demonstrated the strikers' solidarity. Or did it? How do you tell apart those who mock an enemy from those who are disguising a secret and cowardly strategy? I was laughing too, let me tell you. Laughter is an imprecation. Through the open mouth the devil enters and does his work.

Over one hundred men who had attended that meeting returned to work the next day, lured by Elusen's amnesty. Sal was among them. The Gravanos and I were climbing the mountain path together with an eye to joining the pickets at the quarry gates. It was early in the morning and a mist still clung to the ground, unwilling to let go. The sky was royal blue inlaid with bloody streaks, like a marbled ceiling of a Tuscan cathedral. There was a snap in the air. Sal suddenly stopped walking as we were just short of the gates, leant his butt against a granite outcrop and declared his intention to return to work. It was not premeditated at all. His sons' mouths dropped open. 'How many year you think left in me, eh?'

'A lot, Father.' Paul tried to hold his panic at bay.

'When they go on strike at first I say na-thing. But now I say no. Is when the quarry open again they will punish everybody who strike. Paulo, you and your brother can go work someplace else. You are young men. You are free. But I have stay here with your mother and little sister. How you think we live without work, eh?' Their shock registered in multiple echoes around the mountains. Sal pressed his fingers to his lips, illustrating to his sons that they were shouting. 'I never accepted by this community anyhow. Makes no difference.' He flailed his arms as though Jacob and Paul were attempting to tie him up in rope. 'Don't try and stop me. I leave these days happy here in my memory.'

Paul and Jacob fell back against the scree that yielded under their weight. Paul's face turned white as Sal hastened through the pickets. For a moment he stalled at the gates to look back. Hope raised itself in the brothers. But Sal was yearning for something beyond our forms, for something he'd left in Sharon, maybe even further back than that. He turned and crossed the picket line, making the second monumental decision in a lifetime. We watched him disappear inside the slate arena.

It was the greatest catastrophe of their lives. Jacob started planning what he was going to say to Sal over the dinner table. Paul simply disowned him. 'He is no longer my father.'

'Don't be silly,' I said. 'You can't divorce your father like a wife.'

'He is nothing but my enemy, anyhow.'

That miserable scene was witnessed by the pickets standing outside the gates, whose silent judgements scoured the earth. Paul and Jacob knew the meaning of that silence. They knew they would have to shoulder their father's sin between them.

The true father of all this malcontent rode up to the quarry. Elusen arrived on a black Arab, a sensationally beautiful horse sixteen hands high, moving through all four gaits as it tackled the uneven mountain switchbacks, its tail sweeping over its back and its eyes taking in that whole mess, that tragic scene. Elusen's horse shaved past me with its head canted and caught my eye even if Elusen didn't. The pickets parted for him and Elusen rode to just inside the gates and sat the horse in the waist-high mist that had risen hardly at all in the past hour and the horse seemed to float like a boat upon a milky sea. Elusen's boots shone a deep brown in the stirrups. Champing, with its neck arched, the horse looked down at the pickets and Elusen looked down at the pickets and they felt the same disgust. An Arab was a faithful horse, known to go all the way for its rider. You could ride one to death. It kept stamping, letting the pickets feel the ground move from its temper.

Elusen crossed his hands on the pommel of his saddle and offered panegyrics to the jittering *cynfon* collected inside the gates, loud enough for us all to hear. 'Your ascendancy in this quarry has been thwarted by élitism, etcetera. As individuals you shall now have democratic opportunities to rise above your station. Mr Manning, pay a gold sovereign to every man at the conclusion of this day, as a bonus on his wages.'

His voice was different on the mountain from how I'd heard it in his house. In the mountain air he sounded thin and shrill. His voice had no bulk and forced his words, kind of *crapping* the English language. His mount took two steps forward, three back, three forward, two sideways, steam jetting out of his nostrils. An animal that took my breath away, worth two dozen quarrymen's salaries.

I turned to Jacob, a dumb animal himself, and said, 'That is a fabulous horse, don't you agree?'

'I don't know much about horses, Aaron. A horse is a

horse. It has four legs, one at each corner, a head at one end and a tail at the other. The head is opposite the tail, except when the animal turns around. Then the tail is opposite the head.'

Elusen's horse stretched its neck to crop grass, its head disappearing in the ground mist. Elusen jerked at the reins, hauling the animal back. He quartered it and kicked in his heels and led the way into the quarry with a train of *cynfon* following.

Around a third of the pickets assembled looked each other in the eye and something snapped in them. In one big rush they entered the gates. A sound of booing followed them. Paul watched helplessly, seeing men's resolve dissipate. The sound of ringing hooves on slate grew weaker and weaker, replaced by the river's pouring. The various departures left a despondent hole into which we all fell.

Stretched out on the mountain path below were the whiskered heads of two thousand men, making their way up to the gates, inexorably, out of habit. In that instant I recognized the real possibility that they might keep on walking, stage a mass return to work, acting with one collective mind. They didn't look as if they had the stamina to fight on. They were burnt-out cases after just a month. I rejoiced in that hope. Things wouldn't be so bad; it would still be slate they'd be extracting. I wanted the strike to crumble because I wanted to close the file. The underlying reason for that was to do with Paul. His presence always produced a queasiness in me. He was the alcohol that muddled my mind. He coloured any pure thought I had.

As the first wave reached us, Paul spotted the same sign in their bloodshot eyes. Fifty yards to our right was the rock from which Gladstone made his address to three thousand quarrymen during the general election campaign of 1868. Paul broke away and hurriedly made for that same twisted heel of granite kicking out of the mountainside. He climbed, feet slipping on the moss, hands fumbling for handles. At the top he straightened his body and spoke out over their heads.

'How lovely on the mountain are the feet of Him. Can you feel His pleasure, gentlemen? For Jesus said, "Upon this rock I shall build my church and the gates of hell shall not prevail against it."' Paul's voice fell like a silk shroud on the mass. I have never understood where such a man, with no formal education, got the confidence to arrest a multitude like that. Where did he find such resources? He was not a preacher nor a politician yet he had prevented a stampede into work.

He allowed himself a dramatic pause that was filled by a shrill and distant shepherd's voice calling his dogs to heel, like a lullaby travelling the miles of dead ground to our ears: 'Hup! hup! hup! There, there, there. All right, all right, all right . . .'

'Until today the closest contact Lord Elusen has had with slate is leaning over his billiard table.' The truth of that made me laugh, too loud I feared, too knowingly. 'He may act as if the quarry is his private property in the same sense that his purse is. But the quarry is the property of the mountain and the mountain belongs to the Creator. He inherited the quarry from his father but he did not inherit the earth. The man who regards God's property as his own is a blasphemer. Those who return to the quarry under Elusen's new terms risk the wrath of God.'

According to *The Times* of London on 4 April 1938: THE
ELUSEN QUARRY DISPUTE MAY NOW BE CONSIDERED AT
AN END AS ABOUT 2,000 HANDS ARE ALREADY AT WORK.

Whoever that hack was must have made his assessment
by looking out the window and counting the heads walking
down Fleet Street. The strike committee recorded the
number who crossed the picket line each day. According to
their calculations 264 men had returned to work by 30 April.

Under skies choked with rain a large mob of men, women
and children poured out the shame. The *cynfon* had finished
for the day and as the first two came walking towards the
padlocked gates someone in the crowd threatened to kill
them. Such threats were unprecedented, even in jest. Jacob
pointed at one of the two blacklegs standing nervously
inside the gates, waiting for their escort. 'I grew up with
that man,' he said. 'He once tried to pick a fight with me
after Sunday school. I got a hold on him ... Well, he hit
himself ... From behind, I got him around the neck, and
he hit himself in the eye when he tried to throw a punch
backwards.' Jacob could see my difficulty imagining the
graphic detail and so reconstructed the fight, backing into
me, pulling my arm around his neck. In slow motion he
punched himself in the eye. He turned around laughing.
'He was the only person I know who could fall *up* stairs.'

'What about the other one?'

'I don't know him. He's from the other side of Sharon.

Different crowd there. Presbyterians.'

Jacob wandered off when his sister Leah came within range, carrying the baby wrapped in a shawl. At home he tolerated her, but in public he was embarrassed by Leah and her bastard. I noted the impact her loose dark brown hair, mouth full of promises and neat little body packed away in a woollen dress, had on men in the crowd. On the women too, who slipped Leah's name into their litanies. She strolled by unnerved, the muscles bulging in her bare calves, the calves of a sojourner.

Leah's distraction apart, the crowd at the gates was giving all that they'd got. When all the *cynfon* emerged and began to move forward down the mountain, a pale and immutable tribe herded by policemen in black uniform, I looked for Sal among them but could not spot him. It was a cold wintry evening. The sky became thunderous and at least one policeman became frisky, anxious that his helmet spike might conduct lightning.

We followed them right down into town. Young boys attached themselves to the procession like a tail on a kite. Ever more people came out of their homes to see the *cynfon* being escorted away. It was an unusual sight to see police protecting anyone in Sharon and people like to see unusual sights.

A lot of these guys who returned to work had been driven out of Sharon by their former neighbours and colleagues. A chartered train waited at the station to take them to Bethania, five clicks west, where Elusen had housed them in temporary accommodation. Steam billowed around their heads as they boarded the train. I could smell gunpowder and tobacco on their skin. Slate dust swirled off their clothes; their pores were engrained with it. The police turned outwards to face the pickets to inhibit them from piling on to the platform.

They were Manchester and Liverpool police and a little rough. When they tried to arrest two women for knocking horseshoes together and for blowing rude noises through seashells — the worst behaviour they could come up with — Sharon's own cop felt he had to restrain the draftees. 'They have not broken any law of this land.'

'What's that policeman's name?' I asked Paul.

'John Brothers. We called him Jonah when we were children.'

'But not any more?'

'John Brothers doesn't let anyone call him Jonah now.'

'Why not?'

'Because it's a childhood name. He thinks it's inappropriate for a man of authority.'

'Is he a good cop?'

'What kind of man becomes a policeman in a utopia?'

'It takes all sorts.'

'Policemen all have something to hide.'

'Every man has something to hide.'

'Speak for yourself, Aaron.'

'I am.'

'Jonah's always lacked natural ability to distinguish right from wrong, ever since I've known him.'

'That's true of a lot of people, Paul. That's why I believe in the law.'

'Except law changes. Natural law's more consistent.'

'Is there not such a thing as injustice in nature? Nature should be challenged once in a while.'

'There's chaos when you do that. No culture can survive individuals acting on their consciences.'

'That's where the police come into their own. They protect the culture.'

'Really, Aaron? Whose culture are these policemen protecting?'

Some minister stood on a platform bench, above the crowd, his black suit absorbing so much light he seemed not made of flesh at all. He spoke from under the rim of a black panama. 'Disappoint these policemen who want you to behave in such a way as to justify their presence here.' The noise abated, allowing the minister almost to whisper, 'These policemen do not have experience of religious communities. They come from the dockside on the Mersey. From fire and brimstone.'

Paul had drifted from my side to be quickly surrounded by his disciples. Since making his maiden speech on Gladstone's rock everyone wanted a piece of him. I felt resentful of them, the way they coveted his attention. Their gain had become my loss. I pushed them out of the way to say something, my hands cupped around his ear. 'The minister's good, but not as good as you.' Then I left him to the pack.

I tried to make my way through the crowd jamming the station platform. In the quotidian collisions I felt skeletal ribs, sharp elbows, atrophied forearms, pierce my flesh. Women's breasts felt hard and flat. I stepped on the toes of children who had been dragged along for the day and they never even protested. It was a listless, dog-hungry bunch out there.

Standing on the edge of things was Leah. Her brothers had left her fending for herself. She looked like the spark that could burn all the threadbare protesters like so much tinder. 'Hey, Leah. How you doing?'

'Hello, Aaron.'

'You know what they say? That every man and woman has a double somewhere. Well, I've seen your double in America . . .'

87

The minister's hoarse voice interrupted me: 'It is something else and not the police which induces us to keep the peace.'

'. . . She was Lithuanian.'

'Who was?'

'Your double. So beautiful I just had to go up to her and touch her hair. And I was like a total stranger. Maybe you and she will swap places in the next world.'

She looked suddenly vexed, as though she couldn't keep up with the running, the way I covered so much ground so quickly. I was about to conclude that Leah was a simple girl when she executed a quality reply. 'That's for the Buddhists to sort out, not me. Besides if I don't leave Sharon in this world I certainly won't be leaving in the next. I wouldn't want to inflict this place on anyone for eternity. Come on, I want to show you something.'

Leah took me off downtown. She wanted to show me the baby's father. 'You know something?' she said on the way. 'It was Paul who was first to see the baby – after the midwife. The baby's father hasn't shown his face in our house once. He crosses the road to avoid me, like a fox. Which is really stupid. I mean, there is not a single person in town who doesn't know who the father is. So what point is there in that, I ask you? I'm not going to demand he change the baby's bum. I'm not going to take his beer money away. But still, you can't expect a boy like Eli to act like a man when his mother still runs his bathwater each night, can you?'

'Eli . . . Is that his name?'

'Yes.'

'Where did you meet him?'

'The usual place, the Blue Boar. He keeps out of there

now. Kate and Barbara say he's a shadow of his former self. The baby has put him off his pint.'

Leah walked ahead of me into a furniture shop. We stood with our backs to the door as Eli, a moustached young man in a cheap suit, was sweet-talking a woman inside. He did not see us until Leah sung out, 'He's quite a local hero is our Eli.' Eli turned his head away from the woman to the door, his smile of seduction still hanging on his face. It dropped like water from a bucket, seeing Leah holding the baby. The woman lifted her wicker shopping basket off the floor and painfully crossed our path to get out. Her departure left five local people in the shop, all of whom turned towards Eli to see what he would do. The shop felt like a Wild West saloon, with Calamity Jane about to fit up this maverick with a lead waistcoat. Eli would have stood there all day if necessary, pretending to be a clothes-horse. Leah eventually challenged him in front of his customers. 'Don't you even want to look at him? He has your eyes, you know.' So then Eli rolled up *his* eyes, trying to escape into his skull. 'One day he's going to want to know who his father is and I'm going to tell him where to go looking. One day he's going to walk in here and punch you on the nose. Now there's a thought for you for the next eighteen years, Eli. Have a nice life.'

Leah walked out of the shop in tears. She was still crying by the time she got home to find Saul P. Howells, her sometime beau, with two other men, sitting on the same side of the table in the Gravanos' house. They were like three young desperadoes in red and yellow kerchiefs, plotting campaigns of active resistance with Paul and Jacob. They were planning to lead a group to the Prince Albert as the pub was closing for the night and wait for the blacklegs to emerge; the Prince Albert being the only pub in Sharon to serve *cynfon*.

Under normal circumstances Rebekah would have shown these men hospitality, but she had retired into the bedroom where Sal was holed up as soon as they arrived. Sal had begun to fear his sons. The only thing his sons feared was losing this fight. The strike had become their crusade of good against evil, Christians pitted against heathens. The strikers advanced themselves as men with God on their side, so there could be no turning back. But, as with any war, both sides were claiming holy dispensation.

From the day Sal stepped across the picket line the Gravano household physically and emotionally cleaved in two. The brothers and I ate at different times from Sal, who walked out of the kitchen as we came through the door. They blanked their father out, simply lowered his image over an edge in their mind. Rebekah, manically depressed, presided over this bifurcation, shuttling with great difficulty – each journey more painful than the last – between the two divisions. She shopped for Sal and for us separately, as though keeping a kosher kitchen. Most shops in town refused to serve the families of blacklegs and Rebekah presented them with a dilemma. Shopkeepers who refused to serve her one day, served her the next. On a few occasions I accompanied her downtown with a wheelbarrow to help carry groceries and saw for myself the complex situation she was in. As we passed through the whispering gauntlet, strikers' wives flecked her with spit. And strikers' mothers wiped it off. Those women drove at least one wife to suicide that I knew of. She lay down in Eynon's coal merchant's yard and drank a bottle of laudanum. An inquest ruled she'd taken her own life while temporarily insane. Testimonials claimed to have overheard her use bad language in the hours leading up to her death, which in that country was sufficient evidence to constitute insanity. I've not met such unsentimental women as those strikers' wives before or since. They carried the very real burden of the strike around with them while their husbands were frenziedly trying to upgrade the strike into a jihad. The women could not feed their children on rhetoric. The strike was hurting everyone and it was the women who felt it for sure, in the gut. They could smell the sickness shipping in. The same women who put up with injustices in the home because their only source of income was their husbands now had rights equal to any man's on the basis that neither earned a dime.

One time as I wheeled the groceries back up the hill, Rebekah confided in me. 'I could have been a rockman, you know. Like Paul. I have that gift.'

'I don't doubt it, ma'm.'

'I get these darks and nothing to employ them on. And the terrible dreams I have, Aaron . . . I've seen Sal and the boys in my darks, bleeding from their wounds.' She stopped for breath and made a quick survey of the mountains. The quarry was under cloud cover. Below the line of cloud, scores of people were climbing up to join the evening pickets at the quarry gates. One by one they disappeared into the cloud. 'This landscape will be destroyed,' she said. 'Only the dream of it shall remain.'

So, anyway, these men were sat at the table in the parlour with no one to offer them hospitality. That is, until Leah and I came up the road. Jacob told her what they needed. Howells watched her with her baby hitched on to her hip, spooning tea into a pot and slicing cherry cake. I liked looking at her myself, but thought she seemed resentful. Perhaps at Howells, for what he may have done or not done to her. Her eyes flared as she darted about the room, laying before each man a cup and saucer, tea plate and cutlery. The men kept silent the whole time. Reaching over me she left the cake tray in the middle of the table and went to the stove to pour the boiling water into the pot. Leaving the tea to stand, she sat in a chair between Paul and me.

Howells asked Jacob to ask her to leave before getting on with the meeting. Leah's face was slightly bowed, her eyes fixed resolutely on Jacob. He said as softly as possible, 'Leah, would you mind . . . ?'

Leah stood up so fast her chair toppled over. As I bent to pick it up I heard the front door slam. We waited to hear her footfalls outside but there were none. Howells looked at Jacob and Jacob just shrugged. A few moments later as Howells was saying, 'Crime and disorder are *not* the natural consequences of collective bargaining,' Leah walked back in.

She caught everyone's eye to make sure we all noticed the hortative smile on her face. An uncomfortable silence returned to the room as she picked up the teapot, waving it around recklessly. 'I forgot to pour your tea, gentlemen.' She went and stood behind the three visitors, holding the pewter pot above their heads. Leah then walked behind them, pouring the tea over their laps. The table erupted with men forced to their feet with the burn.

'You imbecile!' Jacob shouted. 'Maniac!'

Leah casually placed the pot on the range and left. Rebekah made an unprecedented appearance, her eyes full of concern. 'Leah just poured boiling hot tea over us!' Howells cried, clutching his balls.

'Someone should go after her,' she said.

'*She*'s all right! Look at these men,' Jacob complained.

'I'll go.' I left before someone else volunteered. The force of nature storming down the hill right that moment interested me more than anything I'd left behind in the house.

I caught up with Leah as she was slipping through the arched doors of Jerusalem chapel. She felt my presence behind but didn't stop, cared not that I had followed her in. I joined her in the front pew.

'The men are complaining you forgot to sugar their tea.'

'I'm not sweet enough for them?'

'Too sweet, I think.'

'How are they, really?' She tried to suppress a smile.

'Cooling down.'

She lay the baby on his back beside her, then leaned over and kissed me, sucking and biting my upper lip, tongue, chin. I couldn't believe my luck. What am I talking about? Luck didn't come into it. Americans create their own opportunities.

She came up for air saying, 'Nobody has ever kissed me like that.'

'You're kidding me.'

'No, I'm not. Most of the boys I know keep their lips closed.'

'Can't you split them open?'

'Like a block of slate?'

'Haven't you ever kissed an American?'

'Now where would I've had that opportunity?'

'An American kiss is like no other.'

'Maybe I should be the judge of that . . .'

'The American kiss leaves the past behind. It starts over again. It is both innocent and passionate, an art *and* a science. And very opportunistic. I bet this is a first for old Jerusalem chapel.'

'Are you frightened?'

'Are you?'

'It's not God who chases you out of His house, it's the deacons. They want to run me out of Sharon and send my kid to the workhouse. I've known widows to get pregnant and lose the status of widowhood. Then *they* get sent there too. Deacons have broken into homes of backsliders and smashed their furniture.'

'How come they leave you alone?'

'Because they're frightened of Paul. He's not a man they can cross. Paul doesn't approve of me, but he never forgets I'm his sister.'

'The Temperance Society is their thing, isn't it? They aren't all bad.'

'Hah! The temperance cause has been under an eclipse since the day it convened. You know why, don't you? Half the deacons in this town hold shares in the brewery companies. Oh, Aaron, if you think this town is a utopia . . . Not for the women it's not. If the deacons so much as see a woman carrying water on the sabbath they castigate her through the chapel newsletter. Even if it's the toilet bucket. You want a shit on Sunday? You have to tough it out. When are you going back to America?'

'I've no plans yet.'

'You must be mad to stay in this place.'

'I like it here.'

'The first feeling I remember having as a child was boredom. I'm still waiting for it to pass.'

'I'll go back home one day, I guess.'

'One day, one day. If I count the number of one days I've dreamed about . . . Grains of sand.'

'You're pining for a world never chanced, Leah. America's no picnic for the women either.'

'All there is here is slate, slate and more slate. It might give the men a thrill but it's ballast in my life.'

'What would you do someplace else?'

'I could be someone eventually.'

'It's tough everywhere you go, Leah.'

'If I was born a man, Aaron, I'd be out of here by now. I just can't breathe.'

'With so much fresh air?'

'Amazing, isn't it? Asphyxiating on mountain air.'

Leah made another assault on me. This time she wandered further down from my mouth. But the chapel did inhibit me. At least that was my excuse and I killed her bid. 'So tell me more about Paul. What's he think about the baby?'

'He's an albatross around my neck. Paul, I mean, not the baby. Since I started courting, all my boyfriends had to take his test of moral fibre. Were they of strong conviction like him? But how can boys have any strong convictions? And is that a good thing anyway?'

'I saw him smoke Howells out with a stare that time you brought him home.'

'Because of Paul I've lost Barry, the doctor's son. He was always trying to ingratiate himself with the labourers by drinking them under the table. He staggered into the house after one epic and that was the end of that. Paul frightened him off. Gary went out on account of his effeminacy, Roy when he enlisted in the Royal Welch Fusiliers and Bill was too handsome.'

'That's no sin!'

'True, and I stood up for myself on that account. Paul climbed down and waited for another opening. He saw Bill go into Penelope Bailey's house, this divorced woman from Chester, and without telling me what it was about dragged me into town. As we were passing her house he slammed his fist on her window. Bill's face appeared between the curtains like a criminal in the dock.'

'What about Sal? Where does he figure in all this? He's your father, not Paul.'

'I could stay out all night and he wouldn't know. I could be doing it under his nose and he would think I was exercising. It's Paul I have to answer to.' Leah straightened her skirt, crossed her legs and pulled them underneath her on the pew. She stared less at the image of Christ watching us from His Cross, than at the misty background behind Him. 'The very first boyfriend I had would've been perfect. We would've grown together. Every time I pass him in town with Mary Pierce I want to cry.'

'You're still very young, Leah. Hell, you're still a teenager.'

'When I reached eighteen Paul accused me of getting old. Old and dry. I couldn't believe his cheek. Our mother was eighteen when she gave birth to him, he said. I asserted myself with him first. Then I began asserting myself in town. Come with me. I want to show you something.'

She took my hand and led me into a back room of the chapel. There was a radio in the corner underneath a cloth of red velvet. She slipped the velvet and turned the radio on. We waited for the valves to warm up. Then suddenly there was American big band music heating the room. 'Hey! That's Louis Armstrong,' I said, rather pleased. The song was 'Laughing Louie' in which the whole band sounded as if they were drunk. Leah swung her arms around my neck and we started to dance around the floor. Armstrong lingered on the notes with a pianissimo effect, and we heard our own footwork scraping the slate flagstones.

'You like music, Aaron, don't you?'

'I prefer a ballad if anything. I like melodies, you know
... Can't take the story out of music. It just collapses into a
heap. How can you tell I like music?'

'Because you know how to dance.'

'Sure, I like music. I used to go to all the clubs on 52nd
Street between Fifth and Sixth Avenues ... In New York.
I've seen Billie Holiday at the Onyx, then gone next door
to Jimmy Ryan's to hear Coleman Hawkins on his saxo-
phone. And I've seen Basie when Holiday was with him.
They dressed her up as Aunt Jemima but she don't like
that stuff. They didn't last very long together.'

'I come in here all the time. Paul won't allow music in
the house.'

'This is devil's music, that's why. You ever hear Lady
Day? She's the only one I've heard who can express simple
lyrics without patronizing them. Minimalizes a song and
makes it fresh. Some singers get in between the song and
the audience, messing it up with effects. One time in the
Onyx I heard her sing "The Man I Love" and, Jeez, did
she ever freeze me in my seat.'

We came out of the chapel to find the Prince Albert pub
under siege. The committee men with the tea-stained pants
were leading the vigilantes. Several *cynfon* emerged from the
pub, drunk and wobbly as mattresses, holding small children in front of them. Paul, Howells and the others had to face
off against a human shield of kids. The Anglican vicar was
also there, fighting the blacklegs' corner. He beat time with
a stick and a saucepan as the children sang 'The Streets
are Free to All'. Their noise eventually brought two draftee
policemen out of their billets. I saw Paul step forward and
announce to the *cynfon*, 'You have no right to come out of

your houses. You have no right to show yourself in the street at all,' before he was struck several times with police truncheons.

3 NOVEMBER 1959

The holiday camp where Glanmor works is situated in a former airforce base, four miles outside Sharon. When it first opened its doors some while ago its owner, Billy Butlin (great name), persuaded five millionaires to appear at the ceremony and lie on the ground while an elephant stepped over their bodies. The animal delicately placed his massive feet in the spaces between them, back and forth, back and forth. Look at us, they said, we entrepreneurs take risks. We turn three-ton elephants into fairies.

Billy Butlin was doing what the first Lord Elusen did two hundred years before him. The former looked at a ramshackle airbase and the latter at the heather blue slate and both saw gold. They were visionaries and neither one from around here. The locals are no good at exploiting their own resources, their visions being primarily religious ones. Butlin's is just another chapter in the history of this place, the cogs of which are turned by foreigners. Is that such a terrible thing? At least Butlin, and Elusen in his time, created scores of jobs for local people.

Glanmor holds down such a job here, as a redcoat. He is one of the arbiters of fun, slinking around in his scarlet suit, pouncing on anyone who cannot raise a smile on demand. I can see why, in Paul's opinion, Butlin's is a meaningless place of business. But there no other work in the area. The quarry has gone for good. What does he expect of his son now that the country has become a tourist objective? At least he's not sucking the moisture out of the ground, eating turf, as Elusen predicted.

If Paul is disappointed in his son, it is because he is disappointed with the age. But this is pointless. We all have to make do with the age we are given. What else can you do? To wish oneself away, into another era, is not just futile it's masochistic. Had Paul been born when Glanmor was, would he not have laboured in a holiday camp as Glanmor would have laboured in the quarry in his day?

Butlin's is a New Deal in vacations: a week's holiday for a week's pay, a campful of fun, a ghetto of laughs. An airforce base by the sea gets a new lease of life. What harm is there in that?

Glanmor and I walk the four miles in, arriving before breakfast. Butlin's at this hour of the morning is like a Magritte painting. White clouds puff overhead. The helter-skelter, big dipper, chair-lift, make frozen statues. Campers stroll about in dressing-gowns, walking off hangovers, night-mares, daymares, and all sorts of history.

The campers are billeted in converted airmen's barracks. Glanmor gives me the tour. The barracks were designed to house men only and not with an ear to maintaining the individual's privacy. As we walk between the huts we hear the intimate suckings and scrapings and poppings of people lying inside, their sleeping heads separated from us by the thinnest plywood.

Without officially promoting it Butlin's offers the best in lip-smacking sex, in the heart of a religious mountain region. From the moment they arrive, women offer them-selves in their chalet windows like mannequins in bathing-suits. Men lean against the clapboard walls outside, smoking and combing back their Brylcreemed hair, and make their choices. Within twenty-four hours everyone is sleeping in different chalets from the ones they were assigned.

The dirt poor, the white trash. Live all year in one slum and holiday in another.

In the married quarters, all the distortions of that institution are on display. On paper and in the mouths of priests, marriage offers emotional security, monogamous sex, children. But here in Butlin's it is clear that most people are not emotionally stable. Few are monogamous. Marriage therefore works against human instincts. It's a mug's game, fool's gold. And the kids are made scapegoats. They get it in the neck. They are subjected to prolonged and ritualized public beatings. As early as eight o'clock young children run past us, escaping their chalets in bitter tears, with cigarette burns and bruises flaring on their arms and the backs of their legs. But Butlin's offers them something too, if only a space to hide out for a while, a place to dry their tears on hell-raising rides. I could write songs with some of the threats I hear these parents make towards their kids. First lines, anyway:

- Hit him, Johnny, hit him wi' tha great big pipe.
- Your da's coming back to brain you tonight.
- Thou little sod, thou little twat, cunt, pig, prat.

A walk between the rows of chalets reveals to the voyeur how marriage can appear, after a few years' wear. The women at the stove are wan and creased from heavy smoking, heavy drinking, heavy loads. They smoke up good as they cook, with one eye closed. Rudders of flesh swish back and forth under their forearms as they scrape egg off the griddle. Grandmothers quake indignantly at the immobile lumpy shapes their tattooed sons-in-law make on the beds, stretched out in stockinged feet, reading tabloid newspapers, catching the little heads of children with casually thrown backhand slaps as they move past. Clouds of blue cigarette smoke linger in the air, shifting to the winds of anger, clinging to the handwashed underwear hanging from backs of chairs. Each row of these chalets is named an animal – koala, mink, nightingale, jaguar, hedgehog, gazelle. Very appropriate for a zoo.

Today the camp is awash with a new intake. Glanmor swans the customers inside the dining-room for breakfast and sits with each table for ten minutes. He gets them to smile, gets them to laugh. HE CALMS THEM DOWN. Butlin's may not seem such a big adventure to the more worldly among us, but for the families, factory parties, divorcées taking advantage of the cut-price winter rates, this camp scares them shitless. Whoever they are, Glanmor gets all their names, ruffles the kiddies' hair, cracks a joke. Ten minutes later he moves on, followed by their little cries of panic at the sight of his receding back. At the next table he does the exact same thing, down to the same joke.

There is something about our world today that makes life feel temporary, threatened. It's been an edgy, neurotic ten years. I notice the effect here, the way it manifests itself as a generic need for fun that borders on the insane. We are secure as long as the fun lasts. Now isn't that a tease? Isn't that a worrying concept? That is why Glanmor gives it everything he's got. He does important work here.

Like I say, he goes to the edge to make things work. So do his colleagues, Buster and Frank, who are also his childhood friends. (I'm glad Glan has friends. Friends are a great comfort to the children of melancholic parents.)

Buster maintains the big dipper, oiling the tracks, reducing friction so the kids can get round the rig at maximum speed. He too is a proud redcoat. 'If the wagon comes off the tracks, I fail. If the kids get thrown out of the wagon, I succeed.' Buster is a speed merchant. His permanent limp was gotten from a motorcycle accident while racing on the Isle of Man.

Frank, supervising the tub races, falls into the icy pool with all his clothes on. Accidentally on purpose. Later, he will run over to the office and make a quick change before tumbling back out, orchestrating the laughter.

Frank, Buster and Glanmor change rotas to stay on top of the game. The big dipper, fairground, bingo hall, miniature railway, boating lake, activities hall – they do all these things. An hour apiece. For the first hour this morning Glanmor turns poor man's preacher, calling the numbers in the bingo hall. I sit down and place my bets. Yellow all the twos twenty-two; red two-oh blind twenty; two fat redcoats eighty-eight; blue kelly's eye number one; white five and eight fifty-eight; red on its own number seven; white top-of-the-shop blind ninety.

At ten thirty Glanmor is relieved and we make our way over to the activities hall, an old hangar where Herculeses and Spitfires used to bed. A clutch of single men and women have gathered at the bar inside. Nervous of one another and divided by gender they lean against the membrane of the hangar as though they are trying to push through.

Glanmor coaxes them away from the bar. He knows from experience that the crowd would be quite content just to get a couple of pints down by lunchtime and leave it at that. They'd probably be happy spending their whole week here, to tell the truth, come sun or rain, sipping ale. But a redcoat's job is to take them beyond themselves. Redcoats are their spiritual leaders. They are entrusted with these people's humility.

Glanmor gets a show going in the hall. Everyone buys a fresh round and finds a seat. He asks me to be the judge, then calls on volunteers for the Miss Cheerful, Charming and Chubby contest. Fifteen women clatter into a line out front. 'What you have to look out for is personality,' Glanmor briefs me. 'Who would be the most *fun* to spend the night with. Now, girls. As I come down the line I want each of you to laugh into the microphone for thirty seconds.'

Glanmor embraces the first contestant. He places the microphone close to her mouth. She starts with a fake laugh, which develops into a guffaw in response to the audience's laughter. He knows how this trick works. He has done it many times before. At seventeen years old Glanmor can make any woman laugh. He can make a hangarful of women laugh. He has his father's gift for mesmerizing a crowd.

The next contestant's a live one. She has the face of a tiny animal like a gerbil. 'I can't laugh to order,' she assures him. 'Watch where you're putting your hands!' Glanmor takes his trousers down. 'I can't laugh at that, either. That's tragic!'

Glanmor wrestles her to the ground and sits on top of her face. An old routine that rarely fails. 'If you don't laugh, the judge over there is going to remove some of your clothing.'

'I wanna pee! I wanna pee!'

'Go on then. Someone . . . get me a mop.'

With the next contestant he encounters a very pale and sad face. I can smell her perfume from here, cheap and sweet like boiled sugar. Glanmor puts his arm around her and gags on the perfume. She lays her head on his shoulder and immediately starts to laugh. Her laughter is rapacious but soon transforms into huge, parabolic sobs. The audience still thinks she is laughing, but Glanmor knows better. For women like her the conclusion to a hefty bout of laughter *is* weeping. 'Go on, go on,' Glanmor mouths in her ear, 'take an extra ten seconds.'

I wonder about her, where she comes from. What is a forty-five-year-old woman without children or spouse doing in this former Hercules hangar, lining up with total strangers to humiliate herself? She would weep all day if Glanmor let her. But he doesn't have that kind of time. Her face collapses at his sudden desertion, wobbling on her stilettos on the site of her abandonment. She sluices away tears with a red-painted fingernail. As he goes to the next contestant, she walks like a drunk back into line, like someone emerging from hypnosis. (What happened back there? Where have I been?) She holds out her hands as though searching for a door, a recovery chamber.

We have lunch with Buster, Glan's best friend, a strangely nervous boy, his face pocked with acne and as thin as a voodoo doll. His nose drips from a cold and he keeps wiping it on the back of his sleeve. He doesn't look at me once, threatened by Glanmor's new friend old enough to be his father. Glanmor gets him to open up, asking him to relate the only story of his life, of when he crashed his Matchless 500 into a brick wall in the Isle of Man TT. 'I was laying me bike over, hunnred and ten, hunnred and twenty, burning footrest rubber, then whoomph! Me and the bike part company. A shower of sparks. Concrete garden wall. Hello there, Buster, nice to meet you. Then someone starts kicking me in the back of the head. I reach over me shoulder and catch hold of me own foot!'

Buster leaves our table in the canteen. If it wasn't for that accident, Glanmor tells me, Buster might have gone all the way. He could have been top boy. 'He spent twenty weeks in hospital in Douglas with his leg in traction. Who's your best friend in America?'

'My best friend? Well, now, I guess it has to be Henry Pinkerton.'

'Henry who?'

'I was with Henry when a New York syndicalist took a shot at him in the Brooklyn Jockey club at Gravesend. The bullet clean missed Henry and hit the bartender in the face. Someone was always threatening to kill Henry, which made him fast and furious company to keep. Like that one time a mobster followed him into a restaurant in New Orleans during the Fair Grounds meet. Old Henry saw the guy coming with a piece and punched him over a table. He kicked his pistol away and said, "I had you ruled off down east cause you was corrupting jockeys and owners and I caught you with the goods. Take your punishment like a man, you punk, and stop trying to kill the messenger boy." Dear old Henry. Survived all that only to die while taking the waters in the Nauheim Baths.'

'Was he a gangster or something?'

'No, Henry was one of the good guys. Ever heard of Pinkerton's detective agency? Henry was my superintendent. I used to be a special operative. A darling of the agency.'

'Is that one, or two different jobs?'

I laugh loudly and surprise myself, after so many years holding it back. It feels good to let it out. So late in the day. I feel grateful to this teenage boy and want to tell him more. 'I was an officer of justice.'

'Like a policeman, you mean?'

'Kind of. The state police give inadequate protection in the States. I take on work they can't or won't do. I mean, *used* to. I got to start remembering to talk in the past tense.'

'Did you catch criminals?'

'Can you keep a secret?'

'Yes.'

'I'm running away from something.'

'You kill a man?'

'Say, some company executive has an over-fondness for whores and is embarrassing the company. We kept an eye on him for them. I've escorted nurses to work through the badlands of Detroit. Screened visitors to atomic energy plants. All sorts. Last year I was out on Long Island Sound guarding oyster-beds. Job changed all the time.'

'Sounds all right, that.'

'Times I've had to become invisible, made like to disappear. Then I had to create an imaginary history for myself so as I could work in the field undercover. So no one knew who I really was. That was my job. You still think history is cruel? I can invent one for you that isn't . . .

'It was Pinkerton's who saved Abe Lincoln from being assassinated on his inauguration. Smoked Jessie James out of his house and at a later date chased Butch Cassidy all the way down to Patagonia.'

'How did you get to be a Pinkerton?'

'Well, a Pinkerton must be pure and above reproach. The public safety and the perfect fulfilment of his calling require it. *Required* it.'

'What we do together isn't pure and above reproach.'

'I never took a bribe in all my days with the agency, Glan. Did everything by the book.' My voice has a shrillness to it. 'Besides, it didn't matter if you did have a secret of your own to keep, that you couldn't afford to go public with. A secret like that helped keep you sharp and on track. Was it your first time . . . With me?' He nods. 'I guessed it was.'

'But I've known for some years.'

'Believe me, there is nothing impure about us, Glan. Problem is, I used to think there was.'

'What do you mean?'

'For years I thought I knew exactly who it was trying to undermine democratic institutions. I had no doubt in my mind who the enemy was. Communists and Negro militants. I was pitted against these people. Then a few years ago I got honest with myself. I realized I had more in common with these people than I first thought. We all have grave problems with the law. So I tried to show my support, but they turned me away. A fag! So then I thought maybe it's because the agency has had such a bad time for so long that I can't think straight. The public turned against us well before the war. Pinkerton's were a mercenary army, they said. A dangerous order of spies, preying on social freedom. We were called to account by the Civil Liberties Subcommittee, all sorts of people. LaFollette in '37 called us un-American. Senate stigmatized us as an irresponsible brigade of hired bandits ... Sorry, I'm bending your ear here, Glan.'

'S'all right. I don't mind.'

'You look just like your father.'

'Don't change the subject.'

'Employers still want us, even now. So I don't know what's going on any more. My head's fit to bust. We have operatives in every union in the United States. In United Rubber, United Mine Workers, United Textiles, United Auto Workers. Fifty Pinkerton operatives are working undercover in the auto industry even as we speak. Client is General Motors.'

'Are you on a job now?'

'No, no. I just said why I was here. I've come over to

think things through. Maybe I hope to find some answers here.'

'Can I tell Buster?'

'You said you could keep a secret!'

'I was just asking, that's all.'

'Bottom line is I still have to keep my true identity from being known. I never know who I might bump into from another time and place that I knew when I was on a job that they didn't know about. You see what I'm saying? I am alone in this world. I can never confide in a soul.'

'You confided in me.'

'Yeah, and don't let me regret it.'

It's two o'clock and showtime again. The husband and wife contest: How Well Do You Know Your Spouse? Now I've seen the state of marriage here, I realize why this contest is such great entertainment, a big draw, a prizefight. Yet the questions Glanmor asks are innocuous enough. He does not need the boat to rock more than it already is.

'What side of the bed does your wife sleep on?'

'Right side.'

'Right side, wrong answer. What size bra does she wear?'

'Big cups. Like pudding basins.'

'Who goes to sleep first?'

'*It* goes to sleep first. That's why we don't have children.'

She winces at this last remark, like a baseball has hit her in the face.

'Does she ever wear a see-through nightie or haven't you noticed?'

'You wouldn't want to notice my missus.'

'Over to you now, Missus. What were his first chat-up lines?'

'We didn't start off by talking. And we don't talk now.'

'Does he snore in his sleep?'

'Snore! He pulls the curtains off the windows.'

Evening comes and all the campers want to do is dull themselves into a coma. Butlin's has the biggest pub in the world to accommodate this. Imagine a football stadium with its goalposts removed and you'll get the picture. A wrap-around bar with red chairs and beer-soaked tables hurled into the arena. Women appear in tight dresses advertising more midriff than cleavage. The men steer pints of ale around the floor like lanterns across a moor. They look sore afraid. It is this holiday thing again. They do not know what is expected of them, whether they should be making new friends or staying close to the group they came with. The band on stage is wasted on them. Their mix of bop, blues, is too sophisticated for the average punter. They shake out a cluster of notes that fall upon the audience like glass shrapnel. A party of mongols is dancing in a circle of light. They are the only sober ones here. They are also the only ones who can move rhythmically to the beat. Their intrinsic sense of fun needs no prompting and sets a good example to the others, sets the guidelines for normality.

Glanmor is in there dancing with them, his sun-bleached hair bobbing to the tunes. Watching him absorbed into this handicapped party, I think that if he were my son I'd be proud of the way he relates to all men, all women, has so much to give. He does everything with a passion. So okay, it's easier to sustain passion when the targets keep moving. Passion feeds on revolving, not familiar, flesh. Campers don't die here, not intentionally anyway. Glan gets a fresh intake each week. All shapes and forms. Sometimes nothing appeals maybe, but then he needs his lulls too. Surf's up! is not a call he would want to make every day of his life.

*

I'm reading to Glanmor from a library book. We are down at the beach, sitting on the freezing sand. The book is called *The Surfer Gods of Hawaii*.

'"In the beginning God created heaven and earth. And the earth was without form and void; and darkness was upon the face of the deep. And the spirit of God moved upon the face of the waters." Genesis 1:1–2. Everything has a history, Glan. Everything we do has been done before, even surfing. Here we go. Listen up . . .

"According to official accounts, the first recorded sighting of men riding ocean waves was made in 1778 by the British explorer James Cook in the Hawaiian Islands. He was enamoured of the spectacle of men riding huge waves – as was Mark Twain and Jack London after him. They intuited that surfing was a religious experience, confirmed by the fact that each surfboard was made from sacred trees, the koa or wiliwili.

"This practice was also duly noted but with less enthusiasm by Calvinist missionaries who began arriving in Hawaii after 1820. The missionaries had different ideas of how the Hawaiians should worship. Among the first of the missionaries to come ashore was a man named Hiram Bingham. Passing on his launch through a shoal of surfers waiting to catch waves, Bingham wrote in his journal: 'The appearance of destitution, degradation and barbarism, among the chattering, and almost naked savages, whose heads and feet, and much of their sunburnt skins were bare, was appalling. Some of our number, with gushing tears, turned away from the spectacle.'"'

'Have you seen anybody surf before? Before you saw me?'

'If you lived in Pamplona you'd see a bull or two, wouldn't you? Likewise California. Surfing's number one sport.'

'You didn't tell me you lived in California.'
'I do now. Anyway . . .

"Twenty-seven years after the missionaries came, the Hawaiian population was decimated by European diseases. The population shrank from 300,000 to 40,000. Surfing as a religious expression was practically wiped out. Bingham was ecstatic: 'The decline and discontinuance of the use of the surfboard, as civilization advances,' he said, 'may be accounted for by the increase in modesty, industry, or religion.'"

'He sounds like a Methodist.'

'"A few islanders managed to keep the sport alive. In 1907 a Hawaiian named George Freeth travelled to Southern California to give public surfing demonstrations. They were called aquatic performances, a publicity stunt thought up by Henry Huntington to advertise the newly opened line to the beach of his Pacific Electric Railway."'

'Is that *Huntington* beach? Where they surf under the pier?'

'"Until 1907 surfing was restricted to the islands of Hawaii."'

'Wait a minute,' Glanmor interrupts. 'Hang on a minute. My grandfather had a surfboard *here*, in Sharon, *three years* before one got taken to California.'
'Well, there you are, you see. You *are* interested in history.'
Glanmor tells me how he found Sal's two-hundred-pound surfboard hidden in a hawthorn hedge at the back of his house. It was too heavy for him to lift. But something stirred in him from that moment. He asked Paul what it was for, but Paul didn't want to talk about it.

'That's because your father and his father had a falling out. Real bad one. Worse than anything you or he go through. You probably reminded him of it when you asked him about the surfboard.'

Glanmor showed the surfboard to Buster who suggested hollowing it out.

'Does Buster go surfing with you?'

'Buster? He can't even swim.'

They carved out the surfboard with a chisel, stripped plywood panels from an old door and made a lightweight deck. They waterproofed the joints with pitch. The hull they planed into a gentle curve, sawing off extra pounds from the nose and tail and shaped them into points. At the tail end they attached a twelve-inch fixed keel like a shark's fin and cut a plug-hole for draining water. The conversion reduced its weight by a half. He could now carry it for short periods at a time, balanced on his head.

'Once I started I never wanted to stop. I can't think of anything better to do. Some mornings I'm in the water as early as five. Middle of winter and everything. I know in Dad's day a man found out his true nature in winter. And I know Butlin's isn't quarrying. Butlin's is just a job. But surfing is a test of character, just like he talks about. In winter it is. Definitely. Winter is my time too.'

'Then you are your father's son.'

Glanmor has been surfing for five years; five summers idled away waiting for volatile weather conditions. At the end of a working shift on a hot July or August day he takes his board into the sea and paddles out, distancing himself from the bathers bobbing around in the shallows. He lies on the board, his cheek pressed against the deck, and lets the cool water ripple over his nose and mouth. His mind drifts. The fusion of voices from the beach sounds like balls of cotton wool rubbed together. He capsizes himself every couple of minutes. Last August Bank Holiday he paddled out to where a thirty-eight-foot ketch was anchored in the bay. Its tender was gone so he climbed a ladder off the stern and explored below. There was Kentucky bourbon stashed away in the galley. He drank a mouthful straight from the bottle and in an instant understood how America became a superpower.

America is a source of fascination for him. Primarily because of surfing he feels an affinity for my country more than for his own. Which goes some way to explaining why we hit it off. We are close in heart's desire: he wants to be in America and I want to be in him.

He recorked the Wild Turkey and jumped ship with his granddaddy's surfboard. He paddled towards the coast as Salvatore had done one stormy night in 1904.

SUMMER 1938

After Paul became a politician my private time with him
was much impoverished. He was like an actor who waits
for an audience to appear before revealing anything.
Through the text of course, always through the text. And
the bigger the audience, the greater the revelation. Nor
could he ever relax and let the inevitable happen outside
his door. He held himself responsible for the ways things
were unfolding. He had no filter, no capacity for resting-
up. He was too damn serious, in my opinion, wandering
everywhere in a preoccupied mood. I wandered around
too, preoccupied with him. I could never let him pass for a
moment without following every marvellous effort his cobalt
blue eyes made, or the way his pale chest disappeared
gradually as he distractedly buttoned his shirt in the morning. There was never a fire in the bedroom hearth and I used
to lie under the blankets with all my clothes on and watch
Paul get dressed through the vapour of my frozen breath.
Downstairs I'd place myself in tight corners in the hope
we'd collide as he tried to pass. I mourned whenever he got
by me unseen as I bent down to retrieve a pencil or tie up a
shoelace. When I could stand it no longer, I took Leah to a
secluded location and exhausted my frustration on her.

Walking to the sea, every few minutes we'd stop and kiss
over the baby's head, warming up. The sea was in a playful
mood below the bluff, swollen with surf and kelp. 'What
are women like in America?' Her voice had a ring in it, like
a miniature bell.

'They come from somewhere else originally. All sorts of regimes. America sets them free. They dress how they like, fire guns, smoke cigars and have sex on tables. Sometimes all at the same time.' Leah laughed at that. 'In America all women are acquainted with sex by the time they are sixteen.'

'I was seventeen . . .' Leah said dreamily. She walked ahead along the narrow path bordered by field mallow and wild primroses. Sunlight flowed through her thin summer dress, her legs ghostly behind the muslin, as if not made of flesh at all.

The baby asleep in a shawl by our side, we sat in sheep-nibbled grass on the cliff ledge, drinking in the view of the bay. Leah leant on her hands planted into the ground behind her, holding herself in a posture that looked affected, like something she had seen on another woman and admired from afar. I ran my hands through her bobbed dark hair and her head went limp, her neck rolling and arching. I started removing her underwear.

'You won't think less of me, will you?' she mouthed.

'Of course not.' I felt her taut mountain-girl legs with my hands, kissed the tops of her thighs. Her moans came tumbling out sharp as knives. Grass between her legs tickled my nose. 'Your skin is as soft as a carnation. That's French for complexion.' I opened her blouse and sucked on her nipple. A sweet, watery fluid seeped into my mouth. It tasted quite nasty and I didn't envy the baby at all.

Then I made my less than tender entry; a few seconds later accomplished a thief-like withdrawal.

She sat up looking to see where I had gone, still simpering and mincing, sullen in the mouth, her lips like chalk. I turned over on my side, eyes fixing on the sea below. 'Is that it?' she asked.

'It's the going rate.'

'Eli had amazing endurance.'

'The baby's father . . .'

'The baby does not come from a stork, I can tell you that.'

'Isn't it always the irresponsible men who take their time?'

I picked the baby up in my arms. I didn't like babies as a rule, despite being one myself once. But that child was okay, particularly when he smiled. Every time he smiled he'd fall over. Then he'd pick himself up, smile and fall over again. He was nine months old by now and still had plenty of time on his hands to recover from his mother's adventures.

'Eli was like a bloody piston.'

'Tell me about it . . .'

'The night Jake was conceived was starry and warm. Eli and I walked from the pub to the quarry and made a little nest besides the lake. After a few minutes we removed our clothes and I glimpsed his erection in the moonlight. Never one for words, Eli made his charge straight away. All the foreplay had been taken care of in the pub. But I was ready for him and began to jump about underneath, getting the measure of things, spreading out my chambers, making room for him in there, making him feel comfortable, the furniture shop assistant arranging house. It's funny but I always forget what sex feels like as soon as it's over. Trying to remember the sensation is like trying to remember a dream. But with Eli I felt a connection with all the other men I've had, stitched together as though by a thread. Eli bragged on and on, but what I sensed strongest was the grief of his limited manhood. I tried to imagine my mother underneath my father. Does a life-threatening occupation improve a man's performance? I had all these thoughts, all kinds of things going through my mind. Sex with Eli afforded me a great deal of privacy. He allowed me my own thoughts, you could say that for him. I gave him a tap on the back as he hammered away up there, to remind him that he had company. He kissed my mouth for the first time. But I think he was really only trying to be polite. He formed this really tight seal between his lips and mine and started moving so fast he was leaving me behind. I laced my hands together on his bare bum and dampened his movement until I could feel a connection. Eli and I were two secret lovers in this evangelical town. Here, the rule is, you can only have sex inside marriage, inside the ovulating period and preferably between a hole cut in the sheet. But I felt some shame, let me tell you, even a cynic like me. We were our parents' children after all and there was no escape from that. Eli and Leah – chained to the Scriptures by our names. Eli maintained his two-beat presentation while his mouth kept sliding off my own, making scrambled returns, falling off again. He would sigh in a most affecting manner while promoting contractions and expansions inside me – grand openings and minor disappointments, setbacks, draw-backs and so on. We were like two lumberjacks in the woods either end of the same saw. As my body began to tire I started narrowing the distance between us. I went up to see Eli raking away, approached him about rounding things off. For a second or two I even got to a stage where he began to feel he belonged to me. Then the moment passed and I lost him again. I felt like a flipped coin, spinning between sensation and apathy. Part of this was my own fault, I admit. I would look into his face on the edge of my orgasm, catch sight of a single hair hanging from his nostril and all my passion got dumped through a hole in the floor. A wisp of beery breath had the same effect. But at least Eli had stamina. I didn't feel rushed. After a while he began nuzzling against my breast. He was actually blowing instead of sucking, but then with Eli you had to expect that. I helped him work on his breathing problems until I felt the two parts of my body begin to connect up, as though a telegram was being sent through my flesh. I was close to brimming over, but got tangled up again. Too much pressure against the cervix, against my mouth. I lost the pure line. Things got a little cloudy with all the activity. The obvious target grew remote. One sensation had to go to lighten the load. I plucked his mouth away and hid my face in the night. Immediately I began surging, scaling the heights. Freeing Eli from one of his commitments was the right choice from his point of view too. I left him sucking my nipples hard, like I'd shown him, to chase my own excitement around the quarry. I began to lose track of where I was, who I was. If someone had called out my name right then I doubt I'd have recognized it. Eli began to come at last. It's fabulous how men ejaculate a little before ejaculating the lot a second or two later. That little spurt forewarns a woman that he's on his way. Like a scout sent ahead of the army, trundling in hot from the march. He started to judder and howl and I felt the pressure of warm fluid splash against my insides. A soft bullet hitting the target. A sensation of warmth rolled over me in waves, one after the other, each one reviving the sensitivity of my vulva, numbed for so long by the thrashing furniture salesman. Then everything blacked out. The whole quarry fell and buried us under slate.'

*

They came in on chartered trains, shapely locomotives with long tight cylinders and huge wheels greasy and hot. The strikers who had left for the mines in the south were returning for their summer holidays. The air was crisp and warm and the sky lay wide open, as inviting as a banqueting table. Purple bindweed, sweet briar, bloomed on the mountain slopes. Hope renewed, faith sharpened in their fragrance. The men jumped down from the carriages on to the platform, like potatoes bursting out of a bag. Many bore strange blue scars on their faces and their eyes were dim and smoky. They moved gracelessly, with none of their former delicate balance.

During the previous day sixty-four Dragoon guards arrived from Aldershot, joining the full regiment of soldiers from North Staffordshire. They were billeted in the Anglican church hall and from daybreak saturated the streets. The one question on everybody's lips was, who tipped them off? The homecoming had been a secret among the strike committee up until the day before. Some eyes turned on me, but I steered them away.

As the returning strikers filed through town, they were met by the *cynfon* gathered in small clusters to jeer and spit under the protection of the military. We may have been on the same side but I could not admire them. I wondered what was in their hearts. What did they believe? That they would live for ever? One day the strike would end, while their betrayal would never be forgotten from one generation to the next.

The cavalry took up positions in shop doorways, outside pubs and churches. They were mounted on chestnut crossbreeds, heavily boned geldings and mares, while the few officers had bay saddlehorses. By the way the horses behaved I could tell they'd seen combat somewhere. With military horses as with soldiers you can always tell apart veterans from non-veterans. The horses had that kind of atmosphere about them. They looked sullied and greatly outraged the men of peace.

'Soldiers are the enemy of moral societies,' Paul said grandly.

Without armies there can be no civilization, was my way of thinking. 'I appreciate how you feel, Paul,' I said, 'but this here's pretty mild compared with what I've seen. Back in Pennsylvania, company policemen rode in on railroad cars mounted with Gatling guns, all kinds of stuff . . .'

He wrested his arm away from me, his anger catching me by surprise. 'What point is there in comparing such places?'

'I'm sorry. I just thought . . .'

'There's nothing to be gained comparing the community here with one there. This is the only world I will ever belong to. And it's being transformed.'

'Good things can come out of change too, you know.'

'Only for corrupt societies.'

'Change is a natural thing, Paul. It's organic.'

'Organic? Who was it who ordered these troops to our land, Aaron? Who is it trying to change our society? It's not something we want. It's brought on the whim of one man.'

All my adult life I had moved from place to place. I had no permanent attachment. I was an agent of change, by profession, by *culture*. An opportunist. So why did I feel so lousy, like a child scolded for saying the selfish thing? And I felt as wounded as any child. I turned my back and walked away from him.

The column of strikers marched through town, spear-headed by the full spectrum of womanhood – scrawny,

anaemic mothers carrying children in slings fashioned from
faded aprons; grandmothers in black Sunday dresses and
buckled shoes; harvesters with thick necks, weather-beaten
faces, exposed veins and wind-dried eyes; teenage girls
battened down by shyness and obedience – these women set
the tone for the rest of us who walked through Sharon in
total silence, as in a funeral cortège. The soldiers, astounded
by our self-control, removed their headgear as a mark of
respect.

Close on five thousand men, women and children closed
ranks and perched on a ridge between the quarry and
Sharon. It was thrilling to be there, sharing that same
frame of mind. MPs, union leaders from other industries
in the south and north, joined local ministers on Gladstone's rock and condemned the presence of the military in
Sharon as 'the last argument of Lord Elusen'. A travelling
evangelist, who earlier in the day had been selling shares
in a Liverpool shipping line to colliers, addressed the crowd
as night drew in. His face hovered like a balloon in the
light of a Davy lamp. 'Morality is not measured by the
respect one gives to the man with wide phylacteries. Morality is measured by the effort in helping those who are
worse off. And there will always be someone worse off than
you. As this trouble deepens it will be worth your while to
remember the lesson of the priest and the Levite who
passed by on the other side, the man who had fallen
among thieves.'

People began talking over him. He marshalled little
respect. Mugs of buttermilk and loaves of bread were
passed around on the ground below. Lamps were lit and
the wicks turned down, illuminating the whole mountainside like so many fireflies. Sharon's own Mr Parry ushered
the evangelist off the rock and read from Jeremiah:

'"Therefore thus saith the Lord God of Israel against the pastors that feed my people; Ye have scattered my flock, and driven them away, and have not visited them: behold, I will visit upon you the evil of your doings, saith the Lord. And I will gather the remnant of my flock out of all countries whither I have driven them, and will bring them again to their folds; and they shall be fruitful and increase. And I will set up shepherds over them which shall feed them: and they shall fear no more, nor be dismayed, neither shall they be lacking."

'This is the very word of the Lord.'

I had lost sight of Paul some hours ago. When I next saw him he was climbing the rock to address the five thousand, a ballsy thing to do, electing himself to speak to such a multitude. But Paul would have claimed to be acting as Christ's witness, his voice on loan. There was no ego to keep under control, nothing in it for him.

'Our forefathers built a utopia here in among the mountains, where they could neither see out nor be seen by the rest of the world. We inherited it from them.' His words floated over us, blankets in the cold wind. 'Even though the road to Canaan leads through chaos and pain, we need not fear, for our cause is just. For we are the chosen people.'

'Tell me, Paul, do you really believe in what you say, about this strike being like a holy crusade? That we are the chosen people? On the road to Canaan?'

'Of course.'

'Isn't it a bit, you know, expedient, borrowing from the Scriptures like that?'

'It is not just that I believe in what I say, Aaron. The language of the Scriptures is the language of our everyday lives.'

'Yeah, but the strike is in the here and now. Can't you just say, Stay out. We'll win. He's wrong. We're right?'

'I understand what you're saying. Politicians could easily run a scam here with religious rhetoric. But these quarrymen know me. They know me as well as I know myself. If there were any skeletons in my cupboard they'd know about them. The trouble with professional politicians is that you don't know where they've been before they take up office in Westminster.'

A case was brought against Paul by the police for assault and for possessing firearms. (I had nothing to do with it.) I saw the alleged assault outside the Prince Albert. And what I remember was Paul being bludgeoned by two Manchester cops. As for firearms, that was plainly risible.

Paul was arrested and held in custody until the trial. The strike committee hired a lawyer for him who worked *pro bono*, and whose principal witnesses were the minister of Jerusalem, Mr Parry and the local cop, John (Jonah) Brothers.

Outside the assizes a carnival atmosphere prevailed. A few hundred quarrymen stood listening to a preacher standing on a timber milk crate, conducting business: 'Remember them that are in bonds, as if bound with them; they who suffer adversity, as being yourselves also in the body.' If Paul failed to get acquitted he'd become a martyr and the preacher was attempting to get the men acquainted with that idea.

I was able to get into the courthouse. Walked in free, but wasn't so confident that I'd be able to walk out that way. Only fools and innocents would feel otherwise. This was a place of justice. I felt sick the whole time.

One of the two arresting officers read from his notebook.

'Constable Pickering and meself went t'restrain the accused from assaultin' the fellas leaving Albert pub. The accused shouted out t'others in crowd: "Kill them t'hell boys! Now boys, go for them, the bastards taking me t'prison!"'

Paul's brief was cool efficiency. He called Mr Parry to the stand. 'Mr Parry has walked eleven miles to court today to give evidence. A man sixty-eight years old, your Honour.'

Mr Parry gave Paul the kind of character reference that would have got him a job in a nunnery bathhouse if he'd wanted it.

'If the threats recorded in the officer's notebook sound like written, rather than spoken, dialogue, your Honour, it may have something to do with his lack of fiction-writing talent.'

The charge of possessing firearms was heard next. 'We heard this report, like, ringin' in mountains. An' then we saw the accused coming fast from tha' spot. We found revolver on 'is person. On further questionin' we discovered he were firing revolver in t'river, practisin' armed insurrection.'

John Brothers hardly managed to hold his patience, turning in his evidence. The minister of Jerusalem was unequivocally on Paul's side, the strikers' side, but the only side Brothers was on was the law's. It warmed my heart to hear his testimony. As a sleuth myself I liked to see the law win once in a while. 'If there was any talk of armed insurrection in Sharon I think I'd know more about it than two flatfeet from Manchester. Paul Gravano I have known all my life. There is not an ounce of violence in him.' He turned to address the two police officers. 'Are you going to bring in some man every time you hear a clap of thunder?'

A policeman procured a Smith & Wesson .38. 'Wha's this then, a streak o' lightning?'

Brothers flattened his handlebar moustache. Gingerly he lifted the revolver, then suddenly tossed it to Paul. The two officers flung themselves against the wall. Paul held the handgun by the barrel, like a baby clutching a rattle. It was clear he had never seen a revolver in his life. Brothers casually removed the gun from Paul and handed it to the clerk of the court. 'Please return it to the police officers. They probably have more idea where it came from than Gravano.'

Paul was found not guilty on both charges. I dashed to the front of the courtroom and lifted him on to my shoulders. I walked out to the steps where we faced off against a wall of tossed hats, like a flock of migrating birds. Paul's thighs chafed each side of my neck, but I'd have walked till I dropped if others hadn't demanded a piece of him. He was torn off my shoulders by calloused hands.

I had to wait until we returned home to Sharon before I could get close to him again. He sat at the table in the house, elbows up, surrounded by the inner circle, and by Jacob and me. Leah fed him with condensed milk and white bread, with which he soaked up the milk. 'For seven days I've been sewing mailbags, lads. Gaol made a tailor out of me, but I'd still rather be a quarryman.'

I sat well out on the edge. Paul saw me there and came to me. 'Before this strike I used to believe that family was bigger than politics, bigger than work. The strike has made me realize something. That you must respect some people outside the family at least equally. You are a foreigner, Aaron, and yet you have stayed out with us, while my father has joined the *cynfon*. Do you have you a family, back in America?'

'Yes ... no. It's more aah, you know ... They live in Llewelyn Corners. My home's in Carbon County. A long way away.'

'You came to this town a stranger. But you are part of this family now.'

'Are you serious?'

'When you don't have a sense of humour, like me,' Paul smiled, 'then serious is all you can be.' He touched me on the arm, a touch more profound than any one of Leah's maulings, and with that touch all other people in the room melted away. I felt myself boiled down in size and form by the depth of his stare, all my pain and fear reduced to a crumb. 'What about the art of quarrying slate?' I asked. 'You gonna teach me how?'

'Oh, we'll teach you that all right, the moment we've won this dispute.'

FALL 1938

We went down to the beach at night on the last day of summer. The water was as still as a desert in moonlight, the horizon buckled like overheated metal. The moment we hit the cold sand Paul ran like a spooked horse to the edge of the water. He tugged off his shirt and waded up to his waist. His torso was luminously pale. Suntan lines were drawn across his wrists and neck, his hands and face absorbed by night.

I overcame the shock of immersion to appear by his side. We waded further out until we found a patch of warm water. The cliff from where we had climbed just moments ago shone like an iceberg. Paul's arms carved up bubbles of phosphorescence. 'I wake up in the morning and for the first few seconds I can't breathe. I don't know where I am. I don't know who I am. Those few seconds are what it must be like to have no faith.'

I had such strong feelings for Paul, all sorts of strong feelings, and unloaded them on to Leah. It gave more comfort than anything, feeling her warm skin in some cold slate knoll. After finishing we lay on the ground and watched the stars reflect in the lake below, a sheet of raven black. Leah pulled out burrs from the hem of her skirt and said, 'The slate looks wrung out, don't you think? I'm leaving Sharon, I felt you should know. But don't tell Paul.'

'Where are you going?'

'Paul's rooted here but that doesn't give him the right to chain me to the place. I'm not like him. I'm nomadic.'

'Where will you go?'

'I've been giving a lot of thought to Liverpool.'

'Liverpool?' My only experience of Liverpool was of its policemen drafted into Sharon.

'I went there once with Barbara and Kate, when we were fifteen. We caught the train there and it was dead good. We got adopted by four German sailors – they took us to drinking clubs on Parliament Street. Kate and Barbara stayed with them at their hostel but I caught the last train home like a good girl. Paul was waiting up for me when I got in. He saw a hole in my stockings and cut off all my hair.'

'What about your family, Leah? They're falling apart already. Think what this will do to them. Who's going to look after your brothers ... and Sal ... when Rebekah goes on her travels around the bedroom?'

'What's the matter with you, Aaron? They're all adults you're talking about.' She brushed herself down. 'I've got my life to fulfil. Come with me if you want.'

'I'm too involved here.'

'You're going to get trapped if you're not careful.' Her voice was uninflected. 'The more time goes on, the more you seem one of the faces in the fog.'

'Have some faith in me.'

The next day I woke to the sound of voices clashing downstairs. This was no routine eruption, either. It sounded like a swordfight. I squeezed my eyes shut, making the dark sparkle. Jacob and Paul were already out of their beds. I threw the blankets off, listening the whole time.

Downstairs Leah was sobbing in a heap beside an unlit fire. Hanging kettles above her head were swinging hazardously. The baby sat in a high-chair nibbling fingers of bread while Paul rinsed his mouth out. 'You'll never find your way back here.'

'I don't care. Every day's like a funeral here.'

Paul, a Noah of a guy, kept railing against her. 'This town is being taken apart, but we'll rebuild it. The forces that challenge us now will move on to the next place in the world. You'll walk straight into those forces somewhere else.'

'What are you talking about? I just want some fun.' She snatched a quick glimpse at me, inviting me to go fight her corner. I was ready with the towel if necessary. 'I want to walk through cities by night. I want to use my body in the ancient way.'

The house was so cold we all had little drops of water hanging from our noses, including the baby. It struck me as comic. Paul saw me smile and tore into me with a furious glare.

I decided it was a good time to leave the house and went out to split logs for the fire. After executing a hundred troublesome thoughts with the axe, I walked back indoors, arms laden, to see on the table fried eggs on slices of bread as thick as notebooks. Leah had recovered sufficiently to make breakfast. I touched the egg with the palm of my hand and felt it cold. My tea had a film. It was not my intention to take Leah to task when I asked, 'When did you make this?'

She lifted the wire fireguard to ceiling height. There was a clicking noise as it connected with the lamp. Pieces of thin glass fell on my plate. The baby started to scream. A black Bible passed by my face. It missed Paul by an inch and slithered down the wall. I just caught sight of her vanishing out the door with the baby draped over her arm.

I picked the Bible off the floor, Sal's King James edition. In that fallow period between losing and regaining his religious faith, Sal had read this Bible with an eye to learning English. In the margins he'd logged his positions. At a glance I could tell what places in the world he'd travelled, without the comfort of belief:

> Malachi 1–4: Head sea. Force 2, wind SSe. 12 miles E by S of Harbour Grace, Newfoundland. Carrying 2000 quintals of fish.
>
> Hosea 7–11: Aruba to Rotterdam with phosphate rock. 388 tons. 17°10′N 62°40′W.
>
> Ezekiel 22–48: Lobos d'Afuerra. In port.
>
> 11 Samuel 6–23: 58°10′N 5°51′W. Western Isles. Weather calm. Visibility 50 yards.
>
> Proverbs 14–16, 21, 25: Portage with manure to Martinique, Lesser Antilles.
>
> Jeremiah 6–32: Between Taltal and Tropic of Capricorn. Below sick with scurvy.

Paul sank into an unshakeable moroseness after Leah declared her hand. He regarded all flight as hopeless, confident that Liverpool would be the undoing of her. 'We shall never see her back here again.'

'Liverpool is not so far, Paul. I'll talk to her, if you like.'

'It'll do no good. When my sister makes her mind up there is no changing it. She takes after our father.'

Leah borrowed money from the baby's father's parents, took three days to gather herself, and left. She looked like a refugee in a shawl around her head and a suitcase tied with string. I saw her to the station. The wheels of the train had hardly begun turning before I raced back to comfort Paul.

He had become like a man at a funeral who is not permitted to mourn.

In the days that followed my mind raced in an ecstasy of dangerous movement. With Leah out of town I had nowhere to deploy my heart's desire, no scapegoat for illicit feelings, no alibi. I began to feel myself losing control and took off alone on a train, disembarking at some town where the business of the strike had no meaning. I sent a cable to the Chicago office to let Henry, my superintendent, know how the operation was progressing and to ask him to wire me some cash. 'I am having a romance', I added as a postscript. His reply came burning along the telegraph wires.

> *OP:2374 — You did perfectly right warning the client about the incoming strikers over the summer vacation STOP In your direct intervention I can find no fault STOP But you did wrong stepping out with a local STOP If you are caught while active you risk losing the confidence you have built up with the men STOP Keep a tight ass in there, 2374 so the rest of us backing you up can keep our noses clean STOP Henry*

I returned to Sharon and for the next few days tried to keep my distance from Paul lest I blew all my secrets on the back of the one awkward desire. This had the opposite effect to what I intended. Paul's eyes seemed always to be on me, like searchlights. I started edging over to the more phlegmatic Jacob who never noticed the stars revolving around my head. His gaze bounced off me.

As I sat in the parlour early one evening the door to the house kept opening and closing as various men of the strike committee arrived. Smoke leaked from the stove with every entrance. A beam of light sawed across the floor when the minister, Mr Parry, opened the door. Smoke swirled in the

trapped light. Faces underneath caps, whose eyes I couldn't
see, laughed in the shadows. Saul P. Howells leant his
elbows on the table, grinning at me. What did he know?
What did the others know who came in and leered? One of
them said something directly to me but I never understood
what it was. A curious time lapse kept dogging me; lip
movement and words were unsynchronized. My jaw locked,
unable to answer even the simplest questions, however
unequivocal.

The dining-room ceiling had been enriched by hibiscus and
plumeria sometime during the 1780s. In that same decade
Elusen's great-great-granddaddy, when heir to his title,
made thirty speeches in the House of Commons in defence
of the West Indies Trade. I guessed it must have been his
idea to add the tropical forms. Watching over us at the
table was a life-sized portrait of the Elusen clan: a gent in
plus-fours posed with his rosy-faced wife in her bridal frock
and nine children in the grounds of the house. Abundant
clouds billowed behind their heads. The two youngest
children embraced submissive Irish wolfhounds, while
others kept a grip on the collars of stunned sheep. My host,
as a ten-year-old, relaxed his arm around his younger
brother, an eye cocked knowingly towards the house: one
day that house would be all his.

Elusen jammed a cigar in his mouth and in one grand,
sweeping move struck a match on a Farnese captive support-
ing the slate mantelpiece. A peat fire spluttered in the
hearth behind his back. He was dressed in a well-worn
dinner-jacket. My own, borrowed from the guest wardrobe,
was in much better shape. The maids clattered about
removing empty plates from the table and replenishing
wine glasses. I noticed he liked to soak it up.

Lady Elusen tossed her head around, sprinkling my face with a fine mist. She had come to the table at the start of dinner straight from her bath with her thick chestnut hair wet. It had still not dried by the time we finished dessert. Her voice was seductive, hollow. 'When are we going back to London?' She didn't seem to expect an answer, nor did Elusen furnish her with one. She turned to me instead. 'Where in America do you reside?'

'Pennsylvania, ma'm.'

'That's in Chicago, isn't it?'

'Not quite, ma'm.'

'Please stop calling me ma'm.'

'Excuse me.'

'Are you married?'

'No.'

'Imagine that. Do you prefer blondes, or brunettes like me? I imagine you prefer blondes, tough chap like you.'

'It depends on what's under the hair, ma'm.'

'I've got very little under my hair. I don't feel there's much there at all.' She laughed as if to say, that is what makes me such good fun, as though she had been put on this earth to be gay, to attend as many metropolitan parties as possible before the lights went down. She was incapable of saying anything even mildly serious. In fact, to be serious in her company would have been in bad taste.

I began to notice her form as she relaxed under the influence of wine. Younger than her husband by fifteen years, she was a thick-set woman with large bones. Her pallor looked sensitive to a surfeit of fresh air. The country didn't suit her. It didn't suit me particularly well either, but that was about all we had in common.

Elusen, cigar smoke clouding his face, took the conversation over. 'You look painfully thin, Mr Lewis.'

'I'm all right.'

'You must be careful to feed yourself.'

I laughed. 'I'd stand out a little too obviously if I did.'

'But you must be careful not to get sick.'

'I've got the constitution of a horse.'

'I could arrange victuals for you to pick up.'

'No thank you, sir. I'm into the part right now.'

'You are the first Pinkerton to operate in Great Britain. You are making history, Mr Lewis.'

'That's something, I guess.'

'I'm surprised you haven't deployed operatives here before. Strikes are a big business for your agency, are they not?'

'Jim Farley earned near a million dollars from a single lockout in San Francisco. Mooney and Boland, Gus Thiel – a former Pinkerton cinder dick – are all millionaires. Philadelphia's got seventeen agencies alone.'

'I think it's absolutely essential in our day and age to have your sort helping out.'

'Well, sir, it's not that easy. We had some big trouble years ago and the agency hasn't recovered since. We do bank work now. Jewellers. And the racecourses.'

'I appreciate the help you are giving me, Mr Lewis.'

'You're welcome.'

I gazed deeply at the fine claret in my glass. Lady Elusen yawned and lay her head on the table, her hair tangling with the cutlery. Within seconds she was asleep. Elusen exhaled a volume of cigar smoke. 'What do you make of the quarrymen?'

'The more they suffer, the more they qualify as God's humiliated witnesses. They invest in suffering.'

'Religious humbug! The issue reduces to one question and one question alone: Who owns the quarry?'

Taking a long hard look at him I realized something. The quarry was his inheritance; he did not create the world there. Winning the fight against the men, on the other hand, would make him happy. It would earn him accolades from his fellow industrialists.

'Do you think I'm intractable?'

'I don't have an opinion either way.'

'Tell me, what do they say about me?'

'Do you care?'

'I have the average amount of curiosity.'

'They say you are causing so many to emigrate that it's destroying their culture. They also call you, by the way, a pharaoh who will not let the people go.'

Elusen was drinking glass after glass, a mixture of red wine, port and brandy. The more I told him about himself the more he poured it down. 'If my family hadn't invested in this quarry they wouldn't have a culture. What cool cheek they have.'

'If you attended one of their rallies; if you heard some of their speeches . . .'

'Like King Henry on the eve of Agincourt, you mean? In disguise.'

'I don't know about any King Henry. These men believe they're the chosen people. Which is why I don't think they can be beat, when they're on the strength like this.'

'There are five times as many pubs in Sharon as chapels.'

He made a drunken coded gesture to one of the maids, who quickly brought to the table from a concealed alcove, a Burleighware chamber-pot. While still seated and in front of the maid he took a noisy piss in the pot. His wife woke with the sudden intrusive stench of urine in the air. 'Finally,' she said, 'my hair is dry.'

*

136

A silent maid showed me to my room. We walked past a row of landscape paintings hanging on the balcony walls; of shepherds, harbours, Cumbrian mountains. But no quarry, no industrial scenes at all.

The dark bedroom was made darker by an ebony armoire, hand-painted Chinese wallpaper of parakeets and a purple slate bed. The bed had been carved by quarrymen, who made the headboard to resemble a cemetery headstone. Elusen's wife refused to sleep in it from the day it was delivered. It wasn't restful in her opinion. Accompanying the gift, pinned to the wall above the bed, was a framed letter from the quarrymen's committee: TO LORD ELUSEN, APPLAUDING HIS MUNIFICENCE AND JUDICIOUS LIBERALITY IN THE SHAPE OF CHURCHES, HOSPITALS, MODEL COTTAGES. MAY YOU BE BLESSED BY HOLY LIGHT. For hours I lay awake on that voluminous bed, trying to understand what motivated Elusen in regard to the quarry conflict. This much I knew: for almost two hundred years, until 1868, the heirs of the family held a Tory seat in the House of Commons. That was the year Gladstone came to Sharon and the Liberals won, despite Elusen's gerrymandering; granting life-leases and jobs for life in exchange for votes, stationing his agents conspicuously at the polling-booths so they could write down names of anyone voting against him. In the weeks that followed the election, he fired nine hundred men and evicted them from their homes. But it was Pyrrhic. It was just possible that since 1868 the family had been looking to exact their revenge on the quarrymen.

And there I was, Elusen's spy, his messenger boy, like one of those apoplectic deacons that roamed the town, reporting back to the minister the names of any man they'd seen drunk or visiting another's wife. I sat bolt upright. My head was spinning from too much wine on a shrunken stomach. I sank back into the feather mattress. I'd come to like those people. I sat up once again. I wanted to see a clean fight, that was all. I didn't know what I wanted. I lay down and sighed. Truth was I was sexed-in to that community. It is always sex that brings a man down. I felt a retributive kind of sleep closing in. I lay, entombed in slate with my fists clenched, as parakeets flew off the wall and into my face, closing my eyes with their wings.

On the floor of the great hall were scattered valises, portmanteaus. Before I'd even had breakfast I was ushered into the library by the insouciant butler and found myself in new company. Two grey suits rose from their seats as I entered. Eerie guys, not loose in any way. The cut of their cloth was immaculate. Their cheeks were filled with wind. They even had double chins. Now *that* was something I hadn't seen in a long while down among the strikers.

The room was as cold as a meat locker. Elusen appeared from my blind side and placed his hand on the small of my back, nudging me forward. 'These gentlemen are from London, Mr Lewis.' The men he called government representatives held smiles that were clearly hurting them, their real faces left behind at home as imprints in the bed pillows. Spooks are all born to the job; they don't have childhoods like the rest of us, they skip that part. Secret servicemen the world over share the same icy handshake and hollow stares. A laugh would kill them. Their skin was ceramic pale from smoking too much opium, I'd say, and they both had a fondness for over-shined shoes.

'We're heading an investigation of individuals or groups opposed to official policy. In particular, cases of industrial espionage.'

'Official policy?' I gave them a run. 'Whose official policy?'

'The government's industrial programme.'

'You are in a perfect position to help us out.'

'I have nothing to offer you guys, I'm afraid. I'm not interested in politics.'

'Your interest is neutral, we understand that,' Elusen said. 'But as an English patriot I am duty-bound to pass on the information you give me.'

'You pay for it. It's yours to do with as you please,' I said measuredly.

'Is it true the union has quadrupled its membership since the strike began?' I was asked.

'The lockout has become a bit of a test case for us,' said the other. 'We would like to make a deal with you. In return for information we could furnish you with impeccable references.'

I laughed. 'Some deal, more like a goddamn ambush.'

'I would like you to stay on indefinitely,' Elusen said.

I threw my hat on his leather-bound sofa. 'Look here, I was employed by you to discover one thing, namely the secrets of the slate-quarrying art. Before I got a chance to find out anything, anything at all, you imposed a contract system on the men that my eventual discoveries would have made unnecessary. What I'm saying is, this strike could have been avoided. I'll stay on until the dispute is resolved and not a day longer.' I showed them a smile and the palms of my hands. 'I want to go home one day. Before these callouses become permanent. The old lady's gonna complain if I rip her hose.'

Then I thought, what home? A couple of rented rooms above a gun shop: a kitchen to stash the liquor and a bedroom to drink it in. I shared a bathroom with the gunsmith and ate out every night. Sometimes I ordered in a sandwich if I didn't feel like sitting in a crowded diner. What kind of home was that without a friend to make it sweet? The truth was I had more of a home at the Gravanos'.

I left them tap-dancing in the library and made my way back to town through the woods. On the mountain pass a motor car had come to a halt, a two-seater Wolseley Ten, such an unusual sight that I was drawn to the machine and its occupants. As I came within earshot I heard a voice that sounded like empty tin cans crashing into a bath: 'With a book to read and a friend one loves, one could spend an age in this spot and think it but a day.' I reached the car and proffered my greeting and touched the peak of my cap. Sitting in the driver's seat was a handsome, square-jawed man my own age. His passenger was an older gent of around eighty years, with a strong face crafted over time in manly skills: army or navy at a guess. Suntanned too, with bright eyes canopied under sprouting eyebrows. He took his freckled hand from under a tartan blanket wrapped around his knees and pointed to the *cynfon* climbing like ants over the quarry. 'I presume the quarry employs some three or four thousand?'

'If you said three or four hundred it would be closer to the mark.'

'Allow me to introduce myself. I am Baden-Powell and this is my companion Mr Greeves. I am afraid our motor has decided to sulk at the most inconvenient moment.'

'*General* Baden-Powell?' I blurted out.

'The same.'

'Hot damn! You say your car has broken down, sir? I'll go and see if I can't rustle up some help.'

I felt like a boy as I ran into Sharon (I've seen General

Baden-Powell! I've seen General Baden-Powell!), thumping on doors, scampering on without waiting for replies. I charged into halls and chapels, disturbing Temperance Society meetings, prayer groups, spreading the word that Baden-Powell was stuck on the mountain pass in a motor car. Within half an hour I was retracing my steps up the pass followed by several dozen strikers trundling behind, carrying hemp ropes over their shoulders.

They secured the ropes around the car, saying nothing. Mostly they didn't understand Baden-Powell's English: 'I say, this is too much of an insult on the motor!' They pulled on the ropes like a team of emaciated horses, showing their spines to Baden-Powell and Mr Greeves.

I went on ahead to prepare the townsfolk. I felt a strong desire that they should *perform*, wanted them not to embarrass the General or me with slack manners or defensive xenophobic remarks. The car glided quietly into town and two women brought the visitors tea and homemade buttered bread. A crowd of eighty to a hundred people silently emerged from their houses, dragging their limbs, making one of the largest gatherings I'd seen in a month. The old general became subdued by the spectacle of so many listless people staring blankly at his car. They were raggedy, in frayed clothes that had become too large for their withered bodies, their sunken eyes unable to express much excitement about anything. He whispered something to Greeves as a bow-legged child gripped the door handle and breathed uneasily. 'Which one of you is Baden-Powell?' the kid asked.

The General played a game with him. Pointing to Greeves, he said, 'It's him.'

'No, not me ... Him,' replied Greeves. 'That's your man.'

Jacob arrived and put his head under the hood. He had never been this close to a car before. To his credit, Jacob got the engine going. Mr Greeves and the General, who had not once stepped out of their car, courteously returned the empty tea-cups and plates. As they were saying their farewells, the *cynfon* were descending from the quarry, walking homewards along the escarpment. The image seemed to excite the General. 'Well, yes, indeed. They are *exactly* how we saw them in the Transvaal, on the skyline.'

They drove out of town with dozens of children running after them. I borrowed a bicycle, a Rudge Whiteworth with buckled wheels, and kept up with the car for about a mile before the incline got the better of me. I didn't want the General to leave. I felt we shared a kindred spirit. Baden-Powell was thanking me for a final time until he was interrupted by a man singing on the mountain path. I couldn't see Sal, but recognized the voice. Baden-Powell clasped his hands together. 'Ah, my Lord above,' he swooned. 'What wonderful singers these Welshmen are.'

During the time when the quarry ran full tilt men were bound by the regulation of the working day, stumbling out of bed at dawn, stumbling back an hour or two after dusk. The exigencies of the working week gave little time for reflection, for morbidity. Their one day for contemplation was Sunday. While women shared in the Sunday observances, work was strictly male terrain. Quarrying was a very lowly job, but as long as a man had that work at least he knew he was more than a woman. Now, with that structure all shot, the men were down, they were feeling pretty low-tide. As their union funds got depleted they engaged in women's work, Negro work, any kind of work just to keep mouths fed. Paul took in washing from the tradesmen, slapping their linen against outcrops of rocks on the river bank. Jacob trapped foxes by night for the church parish. Lying in bed I would listen to the faint sound of foxes barking and coughing, warning one another of his presence.

The money I had had wired over from the States was burning a hole in my pocket. Before I could spend it I first had to account for it. I made myself conspicuously absent from the house in Red Lion Road several days in a row, walking into the mountains with a snap bucket and flask, ostensibly to look for work. I idled away the time and pricked my hands on thorns before returning with an account of myself; of how I'd been binding sheaves, and picking potatoes.

Rebekah was making bread in the parlour, in a world of her own, and did not notice me come in. Paul sat at the table reading. I took the seat next to him with my back to Rebekah. Paul closed his book contemplatively when I told him where I'd been. 'It's almost October. What kind of farm was it that picks its potatoes now? This is the time for planting, not picking.'

'I don't know. I'm not a farmer.' But the damage had been done. His expression fractured like a glass dropped on flagstones. 'Look . . . if you don't believe me . . .' I threw a few coins on the table.

'They paid you in cash?'

'How else were they meant to pay me?'

'It's unknown, that's all, for farmers to pay casuals money.'

Jacob walked in with a rabbit he'd trapped and kissed his mother. The sound of his lips on hers bruised the inside of my ear. 'Sal needs dubbin for his boots,' she whispered, so Paul would not hear.

Jacob, whose hearing was impaired from the blasting at the quarry, misheard her. 'Dublin? What's he want to go there for?'

'Dubbin!' she shouted and burst into tears.

I followed Jacob outside, to escape the heat inside. As he skinned the rabbit I felt a correspondence between that vermin and myself. Jacob bored into its belly with a kitchen knife. By the time he'd thrown the offal into the river I was shivering uncontrollably.

That business with Paul would not have gone away by itself. I let Jacob go back indoors alone and went to seek out their minister. As I walked into Jerusalem I heard his voice flow down from the balcony. 'We've got to boost morale, Ted, we can't just let events find their natural course. We must actively intervene . . .'

I moved to the pulpit so they would see me, killing the conversation in the balcony stone dead. The minister was up there with a couple of the strike committee. 'Mr Parry, may I talk to you a minute?'

In the private room at the back of the chapel where Leah and I had danced to Count Basie, I told Parry I was losing my faith.

His whole body sprung forward, completely out of his chair. 'It's this lockout . . . It tests everyone's faith. You are not the first to say this.'

I pulled his attention back on to me and planted the foil, laid down the curve, banking on him being Scripture strong, instinct weak. 'I have some money and I lied to Paul Gravano about how I came by it.'

'How did you come by it?'

'I stole it.'

'Where did you steal it?'

'From here, Mr Parry, from a collection plate.'

144

'How much did you steal?'

'Two shillings.'

'Do you still have the money?'

'Yes.'

'What I want you to do,' he said sombrely, 'is return that money during morning service next Sunday. Return it from where it came, back into a collection plate in addition to your own offering.'

'Yes.'

'The strike is forcing many men to act out of their good characters, Aaron. It's lucky for you that you've shown remorse. Where there is guilt there is always hope.'

'It still can't put right the lie I told Paul, though.'

'I will help heal matters there, if you wish.'

'Please don't tell him that I asked you to. Wait till he comes to you.'

My money was no good in that town. All it had done was get me into a whole lot of bad trouble. From that point on I had to live or die on my wits, along with hundreds of other men in Sharon. I kept out of Paul's way until he met with Parry on strike business. I sat out on the hills and prayed to my god.

Paul returned from his meeting with Parry and came to find me, where I was sitting upstream from the house. He was so full of contrition that he even shared a tip-off about work in the forest.

I knew some about felling trees, which I learnt while operative up in Oregon some years before. I impressed Paul with my skills. For five days we hacked mouths in conifers. I pulled him back as trees toppled and gnashed through the foliage, snapping at us through the mouths, the trunks flinging into the air. We cut logs with a double-action saw and split the logs with mallets and sledges. With the branches we made huge bonfires. We ate meat when a hauler worked his horse too hard and broke its leg. The animal was destroyed and we roasted horse steaks in old oil cans on the fire.

We kept to ourselves during chow, separated from the professional lumberjacks. Paul felt threatened by their presence, by the cold wind of their crudities and sporting foul language. He closed in around me during this time and lay next to me on a bed of pine needles in a tent that we had to share with three other men, who climbed in drunk on homemade hooch and hysterical. I stayed awake long after Paul fell asleep, protecting him, making sure that no man came up in the dark and spoiled him. I was all ready to fight for his soul. I listened in the darkness to his slow breathing for irregularities.

That source of income dried up and the muscles we'd built in the forest soon turned to loose flesh. We ate like birds again, looked like birds. Even my nose lost weight. I had no stamina for anything but sleep. Jacob, Paul and I stayed in bed for as long as possible to preserve body fat and to pass the time. I prayed with the brothers when the town started to lose its children to the infirmary. Then the infirmary got so gummed up Sharon itself became an infirmary. A whole town of children were etiolating from malnutrition. When parents said goodnight to their kids they tried not to make it sound like goodbye.

The whole town seemed as if it were sleepwalking; figures slouched through a desert without energy to say hello, passing each other by like deaf mutes. They drifted into the foothills to trawl the river with nets fashioned out of women's hose, or sat on the river bank with a rod and line while their children nibbled on the dough bait or, worse still, on the maggots. Others ate unripened blackberries and any sheep dung they were lucky enough to find. Old, confused men knelt on the ground and ate grass until they vomited. The people of Sharon were once again sucking moisture out of the earth. Their culture was all shot up.

When the hermit shepherd showed up in town on a humid, sultry day, people feared the worst. He came down off the mountain like a neolithic man wrapped in sackcloth, with tufts of wool screwed to his hair. There was an old faith-healer living on Clyne Road, much maligned by the deacons, who wandered in the mountains collecting herbs for remedies. The shepherd made straight for her place and spoke for the first time in a quarter of a century. He accused her of driving his ewes over a precipice. His sheep were full of maggots, constantly twisting and rolling, scraping their backs against slate posts, driving themselves crazy. Most probably they threw themselves off the edge. Her riposte is a matter of conjecture, but she did let that shepherd cross her threshold and no one saw him leave again.

Elusen had been away in London quite a piece, missing these most mind-warping moments. There was an account in the *Genissen* of one of the late-summer garden parties he hosted at his Hyde Park house. The scribe at the party asked Elusen his thoughts on the continuing strike and he was quoted as saying: 'The good thing to come out of this dispute is a new breed of quarryman. A man just as skilled but less dogmatic and who understands his complete dependency upon me.'

4 NOVEMBER 1959

Glanmor's got seven guys gathered in the hangar for a knobbly knee competition, run along the lines of a beauty contest. The contestants are four buddies from Cornwall on a work outing, three navvies from the same street in Birkenhead. Tough guys, all mortally petrified of . . . Well, what?

'It's all right,' Glanmor calms them and winks at me. 'I'm not here to kill you. The judge can do that.'

He gives each contestant a chance to say a little about himself. To loosen their locked jaws. The same simple questions, democratically distributed.

'Where do you come from?'

'I come from Birkenhead.'

'What's your football team?'

'Liverpool.'

'Where do *you* come from, sunshine?'

'Truro.'

'Have you ever *heard* of football? Never mind. Roll up your trousers to the knee, lads. When you've done that, take off your shirts so the judge can have something else to see in the event of a dead heat.'

For working men they look unfit. The gamut of torsos range from plump to the obese. When one Birkenhead builder refuses orders Glanmor gets me to *forcibly* remove his shirt. I pull his T-shirt off his head and we all see the reason for his coyness. It is a very good reason too. His navy-blue underpants have been pulled up over his enormous beer gut to just below his armpits. The women let rip.

They go sick with laughter. His friends belt it out too, at this amoebic specimen with spiky hair. To hide his embarrassment he crosses his arms, tattooed with spiders and snakes and swords.

I speculate on the power of Glanmor's command. The guy could have refused to take his shirt off. He could have resigned from the contest. But he did as he was told and stands humiliated before fifty witnesses. These men will do anything Glanmor wants.

Intrigued to know what he feels like I knead his gut, then take down his trousers. I want to see the extent of those fabulous underpants. I perform this act without a single objection. You can do what you like in Butlin's with a redcoat chaperone. Any covert sexual pleasure is disguised by the routine, disfigured beyond recognition by the game.

Birkenhead tucks his chin into his neck and laughs bravely at his trousers coiled around his ankles. He waddles back into line in his diaper. The game continues but I stay preoccupied with him. How he touches me! I want to throw a towel into the ring for him.

Glanmor calls each contestant to the PA mike to receive a pinch of snuff off his wrist. The majority have never seen snuff. They do not know what it is. It could be cocaine for all they know, but they take it because Glanmor tells them to. I asked him why he did this and he said it was to give a socially handicapped contestant a moment of clarity.

After each man has inhaled the snuff Glanmor selects five women to go forward and feel their kneecaps. 'What you are looking for is quality meat.'

One of the contestants grabs a judge and simulates fucking her. I die a little just watching him. 'Go on,' Glanmor shouts wearily. 'Fill yer boots, number seven.'

An hour or so after the contest Glanmor and I run into

the gang from Birkenhead in the camp pub, a little worse for wear. 'Do you like fat boys or thin ones?' I ask Glanmor. 'I was thin once, here, during the strike and ever since I've been attracted to fat boys. Fat's a sign of fulfilment.'

The fat guy starts itching himself as he sees us approach. He complains to Glanmor that his chalet has bedbugs. What does he expect from a slum?

'I'm sorry to hear that, sunshine,' says Glanmor.

'I'd like to see the manager, anyroad.'

The manager to whom he refers is Ronnie Rivers, the redcoat immortalized for inventing the Butlin's Complaints Department. Ronnie is almost as synonymous with Butlin's as Billy Butlin himself. People keep on asking to see Ronnie even though he died several years ago.

'We can dig Ronnie up for you,' Glanmor tells Birkenhead, 'but getting him to reply to your complaint is a different matter.'

We split Birkenhead away from his friends and go find Ronnie. On his own the guy is utterly defenceless. Walking through the camp he practically tries to hold our hands.

Glanmor persuades him to take a trip on the chair-lift to the headland. 'Maybe Ronnie's on the beach. What's your name?'

'Arthur.'

'Do you mind heights, Arthur?'

'Me? I'm a bloody fly on a building site, me.'

'Right then, let's go.'

I ruffle Arthur's hair. It's like a haystack.

We share a cage 'for two romantic people' and sit on the same side. As we are being winched out I tell Glanmor that I have never had a woman. Glanmor confesses the same. 'What about you, Arthur?'

'Me neither.'

'Never got your leg over, huh?'

'Well, me sister . . .'

'Family don't count.'

We've got ten minutes going out and ten minutes coming in on this chair-lift. Going out I prepare the groundwork, asking him about his friends, his sexual fantasies. On the return journey (we never found Ronnie), he takes a photograph of Glanmor and me with his Kodak and I take one of him. In the background are the mountains, the slate quarry prominent.

'When did you start the working life, Arthur?'

'At fourteen.'

'Really? As young as that? Who is it that sends us out to work when we're still only children? Same people who get us married off in the same unseemly hurry because if you haven't got a wife you're a pussy, a fag, a nancy boy. And the wife, she don't bother asking for children, she just tells you one day that she's knocked up. Ten years down the road and you've got a screaming hoard of kids and you realize you've been tricked into everything. I love my mother, strike me down dead if I don't, but Jeez, it starts with her. Whose hand is it that rocks the cradle?' I slap him several times on the knee; very hard to confuse him. Tough guy stuff. A gesture of friendship that really hurts. 'Women give us hell, don't they just?' I lay off for a few seconds, then slap him again, further up. He winces and rubs his leg. 'You're a big-hearted guy, Arthur, I don't think women appreciate you. This body . . . looks like it could inflict a lot of damage. How many girls get this far up your leg? I know you prefer the girls, Arthur. So do we. *So do we.* But there's no girls chasing us up here exactly, is there? There's just you and me and this redcoat here.'

I press my hand on to his crotch. I know I've been

backing the right horse when the pants start rising. Cages pass in the opposite direction every minute, their pulleys clanking on the wire. They can't see my face as it's buried between Arthur's thighs, and they can't see Glanmor either, who takes me in the mouth. But every minute I hear the campers' fresh cries of outrage, up here in the sky.

Arthur comes with fifty yards to spare as we pass over the roof of the donkey stables. He nearly blows my head off my shoulders. The wind carries his groans away. He is completely withered like an old dirty apron thrown across a chair.

What a ride this has been for him. In the time it took us to snap those photographs of each other, I have taken away his certainties. Our moment together in the chair-lift will remain in all our memories, indelible as three prints run off the same negative.

I have a general question I'd like to ask. Which organization says, a 'moral queer' is an oxymoron? The American Psychiatric Association, that's who. They call my condition a 'disease' . But tell a healthy man he's diseased for long enough and he'll act as though he is. That's why I plunder, that's my excuse. Enjoy, until the Scarlet Letter is lifted from my chest. I don't care. I do care, however, that Glanmor should not waste his youth paralysed by guilt, squandering that vast capacity he has for love, or misdirecting it. Men are all the more dangerous when they do. I am living proof of that. Homosexuality isn't the evil in the picture, it's the suppression of it that is.

Within the hour Glan and I are back up in the chair-lift with another willing hostage. Nice work when you can get it.

The weak milky sun comes up over the escarpment and

pries wide the blue-eyed slate. We reach the head of the quarry and look down on the rusting hulks of machines on caterpillar tracks. Paul moves his head from side to side as if picking up a radio frequency. He listens to the wind, his face tilting up at the late moon. 'Irish Nationalists hid a weapons cache down there a few years ago.' Paul points to the lake below, spot on. He can still find where everything is without eyes to see. '.303 rifles, 9mm Sten guns, .303 Bren machine-guns, 24,000 rounds of ammunition in bandoliers and 24,000 rounds in cartons.' For a pacifist, Paul seems to know a lot about guns and bullets. 'Special Branch came and put road blocks everywhere. They watched the coast for a week but never caught them.'

Thunder in the distance puts a ceiling over Paul's head. The first drops of rain fall. Together we listen to rain on different surfaces: tinkling on slate, sizzling on grass. On the lake down below it sounds like water coming to the boil. 'Now I can see a little,' he says, the rain opening a window into his memory. Rain is one thing that never changes; the rain and the wind. This rain now falling once fell on Jacob, Sal, Rebekah. Paul can feel the mountains all around him as long as it is raining: the contours, the size of them, if not their heaviness. Everything is light and frothy to a blind man, there is no sense of gravity. 'Where is Jerusalem chapel from here?'

Glanmor turns him round. 'You're facing it.'

'Sometimes I think the quarry venture was cursed from the start. It's fitting to remind ourselves that the first Lord Elusen opposed Abolition at Westminster because his investment in the Triangular Trade financed this quarry. He owned eight hundred acres of Jamaican sugar plantation and six hundred slaves to work it.' He pauses for breath. 'Our employer not only inherited that wealth, but a con-

153

tempt for human life too. The Elusen family were still trading in slavery a hundred years after Abolition.'

On the way down we explore one of the derelict hillside farms, a thick, dry-stone cottage with its roof caved in and moss growing up the walls. In the grass arena inside are sheep skulls, dozens of them, some still cloven with wool. The derelict farm is a monument to the failure of survival. The land here is desert. It does not yield up fruit, just stone. Paul bows his head and sows a prayer into the land for the former inhabitants, resisting a truth that is self-evident to me, that the human race is not the chosen species. We do not come first in God's eyes, not even second. One day we shall all disappear from the face of this earth like those hardy hillside farmers, disappear as other species have vanished before us, leaving only broken stones behind. The mountains have stood sentinel over all this coming and going for the past six hundred million years. Only they get to stay on. The mountains make a mark on all who live here. But we do not make a mark on the mountains.

The High Street in Sharon is deserted. We are the only forms of life to be moving, arm in arm, heads bent against the rain, blind as one another in the downpour. Jerusalem provides salvation from the rain, although Paul would still rather be out in it. 'If it rained indoors,' he says, 'it would fill out the spaces in here, like in the mountains.' Rain is god's gift to a blind man.

Glanmor wanders off to where there is a snooker table behind a wooden partition, leaving me to escort his father down the aisle like a bride on my arm. Paul touches the stone pillars holding up the gallery, the backs of pitch pine pews. At the slate pulpit he feels with his fingertips the inconsistencies in the surface, the indentations, pockmarks, hairline fractures.

Several drenched old men sit dully in the pews. They are recalling the din and clamour of a swollen congregation – the same men who wander around the streets cursing Elusen's name at the tops of their voices, shaking their fists at the mountain, which stares neutrally back. They are still stunned from their failure to secure the future for their children. They have long ago lost pride in themselves, the motivation to better their minds in their recreational time.

I leave Paul to go find Glanmor behind the partition, where some youths are slamming balls with their hands across the torn felt of the snooker table. The cues lie broken on the floor. Then they turn on themselves in acts of unprovoked repressed violence, like colliding snooker balls. A deacon appears from behind an oak door and chases them out, through the chapel, past Paul and out the front door. Paul turns his head after the running feet and quotes something he's read: '"The heavy-jawed deacon of Zion, with his white grocer's apron and hairy nostrils sniffing out corruption". Quite good, don't you think?'

We lead Paul outside now the rain has stopped. We walk as far as the gravel pit and Bethania Presbyterian chapel. I remember something and go take a closer look inside. As I approach I notice the glass windows have been replaced by steel-mesh screens. Nuzzling together on these screens, and mating with a contentment that makes me feel sick, are blankets of blue flies. I push open the door between sand-stone pillars. Inside the chapel are dozens of pigs, scratching their backs on the edges of pews, defecating everywhere. The stench is so bitter, tears spring out of my eyes. All daylight is killed stone dead by the flies, sucking their legs, trying to work out a way to get in to feed on the swine.

Eleven months into the strike and what remained of the population were piteously trying to outrun epidemics of scurvy, rickets, TB . . . only to stumble and fall beneath the tidal wave of sickness. As with all drowning men, they beheld supernatural sights. To my thinking these hallucinations were a kind of protective sport the mind engenders to stave off hunger and illness. Some claims were downright foolish – crows commandeering feed barns, driverless buggies racing down the High Street, dogs speaking in Hebrew – while others were quite inventive.

I was sitting alone in the house one day when Leah's friend Barbara Hills walked through the door. Her eyes shone brightly on account of some mineral deficiency. I drifted out of the chair and slowly raised my hand in greeting. Her two young brothers stood rakishly behind her on the step. 'Is Paul home?' she asked.

'No.'

'Jacob?'

'Everyone's out. What's up?'

Barbara pulled her brothers inside the parlour. 'Daniel and Harold say they've seen the devil.'

'Oh. No big deal then.'

'This is serious.'

She was nervous of the consequences if the boys were telling the truth *or* spinning a yarn. She'd been raising the boys, aged eleven and twelve, since her parents left Sharon in search of work three months ago. There was no man at home to discipline them, which is why she came looking for Paul, to act *in loco parentis*.

'You want me to have a word?'

'It needs a man, I think.' She gave them both a stiff look.

'Sure, no problem. I have a way with kids. Just leave us alone for a while.'

She walked back out the door. I let a moment pass, then the first thing I did was kick Daniel in the balls and punch Harold in the stomach. Left no visible bruises. They both rolled around on the floor crying. 'Now you know I'm not your sister,' I said, 'you'd better tell me what happened up there. And no fucking lies.'

'We were playing a whist drive and this man came up and asked to play a hand,' Daniel stumbled over himself.

'We're not allowed to play cards.'

'I don't give a damn about that. What happened?'

They were dealing hands on a footbridge when a man appeared. Daniel tried to hide the cards, but the stranger had his own deck. He invited them to share a hand of stud poker. If whist was frowned on, poker was downright sinister. The boys knew that much at least, and tried to refuse him. 'But he weren't someone you felt you could say no to,' said Harold. 'Something about him there was.'

He took them to a cave and made a fire from damp leaves, creating a volume of thick smoke. It was in the smoke that Daniel and Harold claimed they saw leaping devilish shapes.

I gave Daniel a kick in the ass. 'It's true!' he howled. 'On my mother's life.'

'All right, take it easy.'

When Paul, Jacob and Rebekah returned home I gave them a general synopsis, and a vague idea of where the

cave was. Rebekah kept the boys under surveillance while I
went out with the Gravano brothers.

After an hour of steady climbing the wind changed
direction from west to north and the air grew colder. We
waded through a river where a rusting steel bedframe was
bolted to slate pillars on either river bank. Its chains,
hanging in the water, were agitated by the wind. The river
was choked with the purple flowers of black horehound, a
rebarbative smell announcing its professed cure for dog
bite. Pink clouds covered the foreground ridge and ghosted
the mountain peaks, looming like huge silent bells, as
soundless as Sharon at that moment. No smoke issued from
any chimney, just a thin long line of doves rose from the
centre of town.

It began raining, tinkling on late foxgloves, darkening
their ochre pigment to a blazing vermillion, a fine rain that
drifted as we drifted. The sucking noise our feet made in
the drenched grass was as intimate a sound as sex. At one
thousand feet land and sky welded together into a globe,
inside which the whole range of human emotion that can
occur, occurred. Following Paul I took readings of his great
shoulders bouncing inside his coat, at the greasy collar and
the back of his neck and turned the idea around in my
head of telling him the truth of my mission. I also turned it
over to kill him there in the rain. It was that close a thing.
Lucky for him, lucky for me, that Jacob was there too.

We found the cave, its walls of green slate running with
water. Jacob sat on his heels and prodded the cold ashes of
a fire. Some of the green leaves had not completely burned
through. He made up another fire. As the flames lit up the
cave and the volume of smoke thickened, little demoniac
figures flittered and writhed in the smoke. Nothing about
this fazed Jacob as it did me. He played his chin with his

forefinger and thumb, before starting to dig around cracks in the rock. A short search later he found what he was looking for: a few carved effigies placed in front of a concave mirror. Paul and I bunched behind him as he explained how the mirror was reflecting distorted images of the effigies into a moving screen of smoke, which the light from the fire illuminated.

We agreed to keep a lid on it, lest it serve as inspiration to others. At the house we found Mr Parry installed. Barbara had brought him over, marshalling some religious muscle on the boys' side. He was praying for the boys' souls, going to some length to condemn the devil's work, 'All the way out east and in other countries besides Britain where I have grave doubts about the human race, such as the cannibalism and famines in Africa, and the United States of America in general, which I have a particular enmity towards for causing these things in the first place.'

The following Saturday a party of twelve strikers witnessed an apparition of Christ at Calvary and opened the account again. They had gone out late at night to steal fruit from Elusen's orchards. Returning via the quarry as the sun was rising, they heard thunder break in a cloudless sky. The apparition appeared in the quarry for several minutes.

They sought Mr Parry's counsel and he sensibly wanted to avoid inciting hysteria and smuggled all twelve men up to the Gravanos' house for interrogation, the logic being that that house was already afflicted. Paul, Jacob and I were rounded up as witnesses.

They stood outside by the woodstack and Parry called them in one by one, his face hovering above a flickering candle on the table.

'What colour was the apparition?'

'Scarlet.'

'What else can you tell me?'

'He had a halo.'

'Had you eaten any kind of wild herb or fungus?'

'No.'

'Did you have a headache at all when you saw it?'

'No.'

'What happened to the surrounding area? Around Christ's body?'

'Disappeared, Mr Parry. Everything in the vicinity. The only ground I could see was through his wounds. Red and white light poured through the holes.'

'White light is the water which cleanses the soul. The red light is His blood shed at Calvary,' Parry said. 'That's what that means.'

'I sniffed His heavenly perfume, Mr Parry.'

'From where the Holy Mother kissed him, no doubt.'

'And I tell you something else, Mr Parry. It was lovely.'

When he was satisfied that they were telling the same story, more or less, Mr Parry rejoiced. 'It is a wondrous thing. A monumental boost to us that He did not appear to the *cynfon*.'

'What do you think?' I asked Jacob, aside.

'Let's go up there in the morning and take a look around.'

We climbed to three thousand feet before dawn, lightheaded from hunger, burning off hundreds of valuable calories. As the sun came up two giant figures were projected in the air, a misty halo around both heads. When Jacob knelt so did one of the apparitions. He stood and the apparition followed. I lifted my arms and the other one did so too.

'What's happening, Jacob?'

'The sun's casting our shadow on a fleece of water vapour.'

'Is this what they saw?'

'They should have seen twelve.'

'I wish Paul could see this. It's beautiful.'

Jacob told me to stare at the sun for as long as I could bear it, then turn my head away, towards a rock. An impression of the sun was cast upon the stone, a yellow ball encircled by red and orange light. In Sharon some hours later all I had to do was imagine a ball of fire for one to appear on the side of a house, in darkened windows.

I was made sick from this stimulation, while Jacob was laid up with a viral infection. It was left to Paul to stagger out and find food, seeking work over twenty miles away.

I took to my bed from inanition. Whenever I tried to stand, my legs threatened to snap like twigs. My ribs stuck through my chest. Blood rushed to my head at the slightest excitement. The bedroom walls were full of suns that I couldn't exorcize. Through the window I saw long grass turn to tropical forests. Shimmering white peaks sailed away like ice floes. And I saw Paul walk home as folk slipped lethargically into chapel for Sunday evening service. He was covering the final furlong over an emptied barley field. A greyhound raced past him with red tipped ears as if bleeding from its extremities. A riderless pony cantered in pursuit of the dog. A dry-stone wall stood between Paul and the town. Lying in a hole in the wall was a naked girl, with a wisp of pubic hair like a lizard's tongue. Paul fell on his knees, thrust his face in the soil and began begging forgiveness for walking on the Sabbath, breaking the Eighth Commandment.

One strange event followed another, each managing to eclipse the last. One night I was woken by a bright light shining into my window. I looked outside and saw a beacon in the old gravel pit. I woke Paul. 'Good Lord,' he said. 'I don't like the look of that.' With a flashlight we walked to the gravel pit. A relaxed and gentle climb was marred by a panicked roar of water rushing underfoot and a sound like string breaking from deep inside the granite. The wind kept throwing itself around, switching direction as though restless and bored. Stars fell away in a clear sky and found the quartz and feldspar that sparkled in the granite. Dogs howled and foxes barked and owls hooted. Paul looked back and at me. 'You scared?'

'Scareder than shit.'

Bethania Presbyterian chapel was situated in the gravel pit, which was formerly a pagan site, a druidic circle. A thread of antagonism was stitched into the fabric of its doctrine. Doubt had always coexisted with faith there. And now, services were suspended altogether by the presbyter, who remembered a man he'd not seen before in the chapel the previous Sunday. He'd worn a bowler hat, in the manner of quarrymen, but with a cock's feather in the band. No one could recall him leaving the chapel after the service.

The presbyter left Sharon seeking help, returning several days later from Cornwall in the company of a senior colleague of the church, a pale and thin cleric in a black suit. Without stopping to rest from his long journey, he went from the station to the blacksmith and borrowed a horse, and rode into the chapel. He came out with a cockerel perched behind him on the horse's back.

That handsome brown cockerel rode on the horse's tail, crowing and unsettling the horse, making it skid on the slate sidewalk. It stuck stubbornly to the horse's hide as we followed on foot along the valley floor, crossing the river where the bedframe hung, its chains clanking in the swollen stream. The river was still clogged with black horehound, its ugly perfume disturbing the horse, who wouldn't respond to the cleric pulling the reins up short. The horse went its own way for a while, wading downstream, sipping the scented water, following the drift of flowers. Black cloud came tumbling over the mountains, bringing instant drenching rain. The clouds passed on and the sun warmed the shivering earth. A tricky azure light overpowered grass and sycamores and birches with a metallic blue coating. Sheep and gulls were infused with blue. A rainbow appeared across the sky and the cock crowed louder and more shrilly than before.

The preacher rode along a goat path over a ridge where the sea could be seen rolling on the other side. He was caught briefly in a shaft of sunlight, silhouetting the black rider against white cloud and the white gelding against purple slate. He disappeared over the ridge, but the image of him remained, albeit reversed, as a white rider on a black horse.

He dismounted at the cliff's edge. He stroked the cock's head, smoothed its ruffled feathers, winning its confidence, then tightened his fist around its neck. The cock put up a ferocious fight as the cleric tied its feet with twine. With a great swinging motion he threw the cock over the edge. We watched it enter the sea with a little splash. The cleric instructed the children in the crowd to compose strongly worded statements on slate tablets and pitch them into the sea in the area where the cock had sunk.

I saw Sal further along on the headland, watching ships sail past the peninsula. His whole body was leaning against the onshore wind. The sea was Sal's church once again. He

163

received guidance there. I wanted to walk over to him, felt the tug of it, but was distracted by a tiny boy of three or four asking if I would write out his message on a piece of slate. 'Sure, kid. What is it?'

'Get thee hence, Satan: for it is written, Thou shalt worship the Lord thy God, and him only shalt thou serve.'

'Very impressive. You make that up?'

He thought about it for a while and said, 'Yes, I did.'

The boy threw his message into the sea with all his strength. Where the slate entered, the sea began to boil and froth. He stayed just long enough to witness it then ran off as fast as he could. I had had enough too. I tightened my coat and hauled over to Sal. I shouted to make myself heard in the wind. 'Is that the devil in the water there?' He ignored me so I grabbed his lapel and shook him. 'Is that the devil in the water? I ask you, as a seaman, for the truth as you know it.'

He spoke slowly as if comforting a fool. 'The backwash it come off the foot of the cliff and dig up the sea. Is where the ebb tide meets the current. Makes for eddies and overfalls and races. The devil he not there. No people in the sea for to destroy.'

For some time Jacob had been making surreptitious plans to emigrate. He finally came out to his brother, justifying his decision on the grounds that there would be no future in Sharon after the strike for anyone. The community was too divided, he said. He wanted out. Paul took that on the chin. He thought he could talk him out of it. 'You won't find a better life somewhere else.'

'No, I doubt it. But I might find work. There is no life at all without work.'

Jacob fixed for himself a working passage on the steam

packet SS *Lucania*, bound for New York at the end of the month.

On the day of Jacob's departure for Liverpool, Paul had to be persuaded to pose for Jacob's photographer friend.

(I still have that photograph. I look like a man on the verge of losing his faith. Standing in front of me are Jacob, Paul, Rebekah, their hair buffeted by wind, ankles deep in grass. Rebekah wears the peach bridal dress she made over twenty years earlier. Her eyes are smoky, unfocused. On one side she is flanked by Jacob and by Paul on the other. Jacob coaches a look of approbation from Paul, who appears to be holding his hands in prayer. The real focal point of the picture is undoubtedly Paul; one's eyes are involuntarily drawn to him. More than a pose, he stands his ground. His smile looks as borrowed as his suit; his eyes removed, as though listening to the shrieking of rock. Around the rim of his bowler hat the mountainscape waves a blue note. An entire ridge has been carved out in misty purple steps and his shoulders support a great heap of slag rock. He is too serious too young, encumbered with his own prophecies.)

With both Leah and Jacob gone, the animosity in the house settled into a silent routine. There is nothing deadlier than routine. At least when people fight they know they are alive. So when the town erupted into violent squalls it felt like a blister bursting. I would have exploded if the town hadn't exploded first.

The train of events began on New Year's Eve when a drunken blackleg left the Prince Albert around ten o'clock and lobbed a stone through the window of William Williams's shoe shop. Since the strike began Williams hadn't sold a single pair. His shop had become a shoe museum. Shoes of all sizes were piled high in the windows, covered in a fine soft dust. The drunk scooped an armful and dragged them on to the sidewalk. Williams was upstairs when he heard the splintering of glass, the expletives and the wailing, and made his way down. He stood in the shop framed by broken glass, with footwear boxes stacked to the ceiling. His first concern was not so much the shoes but the man collapsed on the sidewalk, weeping, surrounded by little children's red sandals. He opened the door, throwing light on to the street and went out to see if he couldn't comfort him. He placed an arm around his shoulders. There could have been a reconciliation built round that image, the beginning of a healing between the two sides. But then more *cynfon* leaving the Prince Albert saw Williams with his arm around their buddy, and took it the wrong way. They dragged William Williams off into the woods at the back of the shop.

I found Williams as they had left him, in a heap on the river bank, his legs plunged into the freezing water. I got him to his feet and hitched his arm around my shoulder. Several of his ribs were broken and his face was badly swollen. People started materializing around us and raised a hue and cry. Within minutes the town was thundering with running feet.

Hundreds of strikers came out in their shirtsleeves, an ever-swelling number of atrophied figures, fuelled by adrenalin, who drove a wedge into the besieged town, scrambling noisily through the network of alleyways, seeking out Williams's assailants. They started ripping up the community with a clinical precision, pulling up paving-stones and prising rocks from dry-stone walls with a newfound strength. I left Williams in the hands of women and joined the mob throwing a zig-zag between churches and pubs, banging on doors, laughing like a crazed child.

Faces appeared at blackened windows, dissolving as stones curved up to meet them. We smashed every pane of glass in the Prince Albert before moving on to the contractor's house, with whom this strike began. His wife's face appeared in the window, framing her perfect agony. The contractor ran out of the front door, flinging himself around on the end of a heavy axe, giving everything he'd got. The axe was snatched away and Saul P. Howells punched him to the ground. Howells stared at his clenched knuckles, exhaling with satisfaction. There followed a moment of watchfulness as the contractor regained consciousness and tried to get up. Then the crowd mauled him over some more, as campanologists in the Anglican church were ringing in the New Year.

The pleasure I got running with that crowd I didn't want to end. It was a liberation to lose myself, to become an accomplice to the collective will, doing all those things the law or my conscience had never allowed me to do, beating on people, breaking windows, torching houses. I checked that the gradient was changing under my feet approaching Red Lion Road. Suddenly I didn't feel so happy any more. As the first column arrived at our house, Paul came out and stood in the doorway, pleading immunity for his father. But the crowd would not let it go.

'You want me to beat my own father?'

I pushed my way through to the front, racing with incomplete ideas. 'Walk away from this, Paul . . .' was all I could manage. Paul swung his head around. He had not recognized my voice and looked at me down miles of murmuring tunnel.

On all sides were morbid faces, annealed by rectitude.

They wanted to see blood on blood, the final sacrifice. Paul found some calm and rational ground for a second. 'If Sal's got to be summarily punished, then I'm going to do it.'

Of all the thousand crusading words that passed his lips it is these few that reverberate the most. I still turn them round in my head as I lie in bed or walk alone, viewing them from all sides like strange stones.

I stepped back. I had no right to make any judgement over this. No one was capable of judging a dog in that climate.

Paul entered the house, into the gloaming. I closed my eyes, plotting out his course through the parlour, up the stairs to Sal's bedroom, where he would be sitting on the bed, slump-backed, his brow wrinkled, his mind chasing a myriad different memories. Paul reappeared with Sal, passive as a kitten. Rebekah was doing all the fighting, raining blows on to Paul's back, until she lost her strength and collapsed.

Paul struck his father with the back of his hand, a great meaty clout. He hit him three more times before Sal sank to his knees. That might have been sufficient for Paul and the crowd, but it was not sufficient for Sal, who brought himself back on to his feet to stare into his son's face, challenging him to strike again. Paul threw a punch and Sal fell. Again he climbed up from the ground, but with greater difficulty, hauling his weight on to one leg, then the other, climbing to full stretch to stare Paul in the face, a real deep downhome look, with his arms hanging at his sides. Paul put his fingers together and swung his knitted fists into Sal's head. Sal fell and lay out on the grass for several seconds, before starting the long and arduous ascent to his feet, his nose and mouth pouring blood. Paul was weeping as he tried to complete the execution, bringing his knee sharply into Sal's face. Sal rolled on to his side, raised his head off the slate path, put an arm out, lifting his torso up, pulling in one leg to support his body and rolled his head over his waist, placing his second hand next to the first and coming up on all fours, shuffling his legs in closer to his body, removing one hand off the ground on to his knee, unsteadily removing the second hand off the ground, wavering, but rising slowly, craning himself up, until he stood his full height, facing off with his son, his face a blood mask. Paul hit him again and this time Sal stayed down.

The tears in Paul's eyes turned into slivers of ice. I turned my head away. The deep blue of the sky was blurred by flaming torches and heavy rain began to fall. It clanked against the tin roof of a pigeon shed, like a bell ringing above the breathing of the crowd. By and by the clouds broke up and I looked at the stars, the big dipper looming above the quarry. This was not trouble I had ever experienced before. The trouble inside me now mirrored exactly the trouble everywhere else. It was cruel out there. And it was cruel in here.

I walked down the hill alone, disgusted with women and with all men.

5 NOVEMBER 1959

The Elusen quarry strike was still being fought through the second generation. Glanmor and his friends had been fighting with the sons of *cynfon* since they could walk. All over town they swined it, in eviscerated chapels, abandoned houses, as well as at the quarry itself. They made bombs from homemade explosives and set them off in the Anglican church, where the *cynfon* had defected. After their battles they escaped into the woods and lay low in hideouts constructed from materials salvaged from the chapels. They sat in their den looking through stained glass to see if anyone would fall in one of their booby traps. Glanmor said he fought for his father, as the logical way of winning his love.

Today is the fifth anniversary of such a battle, the battle of Guy Fawkes, when some antagonists had set alight the thirty-foot bonfire, a day too early, that had taken them months to collect for and build in the pit of the quarry. Glanmor was first to discover the hot ashes on the morning of the fifth. Frank came along a few minutes later, followed by Buster. They kicked over the smouldering ash with their toes, too distraught to talk.

Then Wilkie arrived at the quarry and started to cry. Wilkie was Saul P. Howells's son. It was Glanmor who told me what became of Howells, of how he and his wife and Wilkie were driving along the mountain pass when the car left the road. The theory goes that Howells pulled the steering-wheel. They rolled five hundred yards to the bottom of a ravine. His mother and father died but Wilkie crawled out of the car alive. He went to live with his grandmother, who was so chronically depressed by the deaths she stayed in bed all day inhaling her daughter's perfume bottles. Rather than cherish Wilkie as the sole survivor of her family, she punished him for the crime of survival. As a matricide himself, Glanmor might have been more sympathetic, but it didn't work out that way; children are cruel sportsmen. They gave Wilkie a hard time precisely because he was orphaned. All their parents were out to lunch in some way. Religious fanatics, drunks, cripples, burn-out cases, but none of them as *dead* as Wilkie's. Having Wilkie around made the others feel good. He made them feel lucky. The way they showed their appreciation was by stealing off him, pretending not to hear what he said, hiding from him in the woods, baiting his dog. But he never gave up on them. His need for friends outweighed any amount of pride.

The only creature who unconditionally loved Wilkie was his dog. But even that didn't last. They were on a mountain jaunt one Saturday. Wilkie brought his dog along, half labrador, half setter, all rebel. They hadn't been out more than an hour when the dog broke loose and disappeared over a ridge. As they were whistling to try to bring him back, a gun went off. Then Wilkie was away, vaulting over streams, sliding on wet gorse, running full pelt down steep outcrops of rock with no regard for himself. They caught up with him twenty minutes later, kneeling in an intense arena of fresh grass, shaking, holding his limp animal in his arms like a baby. Standing over Wilkie was a bellicose farmer with a broken shotgun. His face was a study in containment as he said something about the dog running off the fat of his sheep. The farmer drove the boys back to town in his Land Rover. No one said a word the whole journey. Wilkie held the corpse on his lap and wouldn't let anyone touch it. The dog's whole face was missing. At his house, Wilkie climbed out of the Land Rover and staggered indoors with the corpse on his shoulders. They might have advised him not to show the remains to his grandmother, but Wilkie was beyond advice. He couldn't be reached.

He kept away for a couple of weeks after that incident. When he turned up again at Glanmor's back door, it was to challenge him out to a fist fight, claiming he'd thrown his school sandwiches into a drain. Glanmor had no idea what he was talking about. He presumed Wilkie was just trying to make contact again. He walked away from this provocation and into town. Wilkie followed him all the way, snapping at his heels until he'd worked whatever it was out of his system.

So, anyway, some punks had destroyed their Guy Fawkes fire. They knew who they were, actually. Wordlessly, Buster began laying out on the ground contents of his school satchel: a tin of sodium chloride weedkiller, a jam jar of sugar, two lengths of copper pipe, a tin of calcium carbide crystals, an empty dandelion and burdock soda pop bottle, a bag of glass marbles, assorted fireworks and a small wooden box of fuses he found under a rockfall in the quarry. He mixed a compound of sugar and weedkiller and filled one of the pipes, hammering down each end with a slate brick. In the way of the Victorian quarrymen he measured out a fuse, three minutes an arm's length, inserting one end of the fuse in a pre-drilled hole. Into the pipe fashioned as a mortar gun, Buster poured a little of the firework gunpowder, chased down by a marble the same diameter as the tube. Lastly he dropped the calcium carbide crystals into the bottle, added lake water and sealed it.

The first contractor to be employed by Elusen, back in

'37, had five sons. Glanmor and his friends had been at war with them since he could remember. They lived together in a house that was surrounded by high fences and padlocked gates. The contractor was an old man now and rich enough to buy expensive foreign cars. On 5 November his Alfa Romeo was sitting in the sunshine outside his house. In a coordinated operation Buster placed the lead pipe in the trunk and lit the fuse. Frank placed the soda pop bottle in front of the gates. They joined Wilkie and Glanmor holding down positions behind the cemetery wall.

The bomb went off inside the car. It was just too damn loud. The explosion was too great for small boys. That was adult noise coming from the car, war noise. Buster had overdone it with the weedkiller. But if they felt overstretched they never articulated it to one another.

The explosion conjured the contractor's sons out of their house. They ran down the drive and stood their side of the gates with heads and hands protruding through the fancy wrought iron work. Buster lit the short fuse on the mortar, his marksmanship under no doubt after he'd shot the stiletto heel off a *cynfon*'s widow's shoe, as she was walking home from a dance club. Perhaps it was more luck than judgement, but his friends were so grateful he hadn't actually killed her that they gave him the benefit of the doubt.

The charge ignited and sent a marble whistling out the end of the pipe, hitting the bottle at the gates. The calcium carbide crystals had been reacting to the water all this time, building up a huge compression of acetylene gas. When the bottle smashed, glass shrapnel flew everywhere. The enemy screamed in pain, with pieces of glass embedded in their flesh.

Severe, but so was their fathers' despair. These boys were

out to avenge them. Glanmor had been so much in need of his father's esteem it was going to take just less than murder to earn it.

They ran to the tree den in the woods, with Buster shouting, 'Good kill! Good kill!' Glanmor's cheeks flapped against his jaw as the nervous howls of laughter left his mouth.

They climbed into the tree, pulling the rope-ladder up behind. For R & R they shared another of Buster's inventions, a strip of fur cut from the bottom of his mother's coat and stitched into a cylinder. They took turns to spit inside and penetrate the wet recess with two fingers, jerking them in and out, shouting, 'Whoah!' inspirationally, discussing the difference between shagging and fucking. It wasn't just semiotics. Shagging was a straightforward thing, while fucking meant taking a piss at the same time. Or maybe fucking meant taking it in the ass? They couldn't resolve the argument one way or the other and turned on themselves in confused fury.

Glanmor got home later that day to discover that news of their mission had preceded him. The house was in disarray. A plate had been broken and lay in pieces where it fell. There was an electricity in the air, a sharp pungent smell of spent anger. Glanmor packed a wedge of cake and made for the hills.

He banged on through open ground, the gradient heavy-going. After a while he sat on a rock and looked back. Sharon was bowed low in the valley. In the middle distance a figure was winding his way up the mountain, tracing Glanmor's path. He moved off, his hands assisting his feet to gain altitude. The higher he climbed the thinner the air and the harder it was to breathe. He allowed himself to cry a little. At twelve years old he was too young for coping with such things. His calf muscles trembled and his throat grew dry. He was ravenously hungry.

With partial sight Paul was navigating the mountain two hundred yards away, his head nodding with every step, resolutely plodding on. Glanmor sat down and stuffed his mouth with cake.

It was an animal trail they were using and he wondered how Paul fitted in. He imagined him catching up, having lost all reason along the way, and devouring him with his strong teeth. He ate every piece of that cake, including some of the greaseproof paper it came wrapped in. Then he couldn't breathe at all. Cake stuck in his throat as his father closed in. This was Paul's terrain. He knew how to make distance without ever getting tired.

Glanmor fell beside a stream and cupped his hands under the running water. He opened his mouth to drink and raisins cased in dough fell out. He tried to scramble the morsels back from the water, as if that food was his only chance, the very stuff of survival. Paul was close enough for him to see his great club-like hands swinging by his side. Helplessly, he watched scraps of cake taken down by the stream, knowing they would give away his position as they passed his father in the water.

Glanmor scrambled on, the going getting steeper all the time. Every so often he had to detour around vertical drops. He crossed the snowline around dusk. Seen from a distance this is a line, but up there, actually in it, the snow lay in patches. He felt as if he were wading into the receding hairline of an old man, stepping through the years of his life back into his youth as the snow thickened. Then it was all snow, up to his knees, and he forgot about images of old men and worried about *his* youth. He used his hands to pull out of the snow and his fingers fizzled with cold. He sobbed as he climbed higher, Paul's closing presence promising a heat he couldn't use.

He had not looked back for several minutes. In the distance he saw a cloud bank rolling in, covering the lower slopes and enveloping his father. Glanmor could hear his voice, like a bee trapped in a jar. He sat down in the snow and waited for the cloud to catch up. The cloud was full of wet mist and saved him from his father. The space immediately in front of his eyes performed tricks. The air filled with needles, turning through their lengths into sharp points and embroidering small blue patterns into the sky. He looked right and left. There was no more mountain. The sky was welded to the snow and his legs had disappeared at the knee. Sweat he'd earned through climbing iced over his skin. It was exotically cold and slowed him right down until all he felt like doing was lying in the snow and resting, maybe sleep a little.

He hit upon a better plan, to head back down. He could pass within an arm's length of his father and he wouldn't know. He descended very slowly at first, laboriously placing one foot in front of the other, the snow holding his back leg, reluctant to let him pass on, to take his next step. He thought about the sheer drops he passed on the way up that he would not see before falling. He plunged into snow up to his waist, ejecting from the hole by the force of gravity, pitched head first. Sliding on his stomach, blood rushing to his face, he put his hands out to brake but could get no purchase. His body spiralled round and stopped. It was then he realized he'd lost one of his shoes.

Losing his father he had also lost himself. The cold destroyed all sense of direction. Gravity was all he had. He retarded his descent, proceeding as though with his eyes shut. He began to feel the snow diminish, marginally at first. He punched a hole into the snow and felt the grass. Dark patches appeared in the whiteness. The temperature grew warmer. The cloud peeled back to reveal good tidings from the stars. He laughed. He felt the grace of God on his side. He had been scheduled to die and now he was free. The moon outlined Sharon, a lively silhouette. He telegraphed the good news to Paul out somewhere on the same flank. Paul trusted in miracles. He would surely grant Glanmor amnesty.

He reached the house and walked in through the front door as Paul was entering the same moment by the back, through the pantry. His clothes, like his son's, were soaked through. Glanmor smiled at him.

Paul moved through the air and hauled Glanmor over the table, unbuckling his belt in one deft move. Glanmor was still so cold that when the belt began to connect he didn't feel a thing. He cried not from pain, but from humiliation. Weren't they meant to be on the same side? He had survived the most unconscionable elements, he deserved better than this. Tears burned the skin on his cheeks.

SPRING 1939

One cool morning when the mountains were noisy with the shrieking of birds, an evangelist came through town and healed three men of their lumbago, rheumatism and diabetes. Hearing of the miracles people stepped uneasily out of the dank houses, blinking into the light. They climbed into the hills on unsteady limbs to see the signs and wonders. The preacher pointed to strangers and called out their names. Individuals spoke in tongues, their wailing voices rising and falling in waves along an indifferent shoulder of mountain.

All over, spontaneous conversions occurred. People were struck down while walking the dog, pinning the hem of a dress, in the middle of sex. Two elderly sisters died from shock.

No external order was imposed upon the meetings held in open country. They continued right through the day until midnight, then started again the following morning. The evangelist just stood by, allowing matters to take their own course. Congregations followed bursts of song by long silences, by song. Sometimes he deserted his flock altogether to go get a meal in town. Now the world had accepted the faith, leadership had became extraneous. The preacher's job was done.

I was in need of spiritual rebuilding as much as anyone, *more* than anyone. I wanted to be taken up in the drama of it all, wanted the spirit to inhabit me. I had to go and see for myself.

What I saw was a slow dance under a purple sky. Shoals of people swayed from side to side, swinging their arms above their heads, waving red flags embroidered with JESUS. I held hands with total strangers and yielded to the power of reconstruction.

The evangelist had a hybrid accent, not unlike mine. He too had worn down some shoe leather in his time. He was dressed in a brown suit with a daffodil in his lapel. 'Jesus said, "I am the light of the world, *of* the world. The light of life." Isn't that wonderful? That the light of the world was a man willing to die for us?' He stopped, as though his words needed digesting like a meaty dish, as if what he'd said was more complex than it actually sounded. I waited impatiently. 'All other faiths are invalid as ways to God. For why else did Jesus say, "No one comes to the Father except by me"?' He stopped preaching and joined in with the silence. I couldn't help feeling short-changed. I didn't think that was part of his job, to clam up. We were there to listen. He was there to preach. I waited for him to tie up the strands, but all he managed was something about revelation. 'Revelation . . . Implementation.' (I missed the bit in the middle.) 'Can you repeat that back to me? Revelation without implementation . . .'

The evangelist's theology was thin on the ground. Tautology was all he'd got. Religion without politics is ethereal. Paul put on better shows than that.

I checked out early. People chanted like clones behind my back.

While there were many in Sharon who fell under the salvationist's spell, for others salvation had come too late. When the evangelist returned the following day he came with his circus, his ministry team, bragging of their intent to heal the rift between strikers and *cynfon*. He was met on the road by a crowd shaking bags of slate on the end of poles to frighten his horse. Mud was thrown in his face. He was pulled to the ground and taken into the Victoria Arms, the one pub defiantly resisting the revivalists' call to shut tap. Liquor was forced down his throat until he vomited. But the sicker he got the more he rejoiced, happy to be deemed worthy of persecution for the sake of the Gospel.

The revival may not have been successful among Sharon's quarrymen, but it did have some effect on the women. They had suffered as much as anyone. While men attempted to bury the trauma of the past two years, women would not let it go. They mourned publicly for a fallen generation. The preacher turned them all into abstract forms focused on some nirvana. Rebekah was one such woman who fell in headlong with the experiment. She persuaded Paul and me to accompany her to a rally deep in mountain territory.

A shelf of rock formed a natural launch into a lake, where a ministry team was immersing bodies, each man and woman screaming and bucking as they were lowered into the cold, soft water. The preacher shouted draconian predictions all over the place, the wind buffeting his words about. 'This kingdom is descending into CHAOS. We shall see an increase of OCCULT practices, ABORTION. This is Satan's work. For the moment a man has conceived a child in a woman's belly, a new immortal soul appears in heaven. That child is already upon the earth.'

When the 'spiritually infirm' were invited to approach the lakeside to be healed in prayer, Rebekah pushed forward, hoping to get a cure for her depression, her darks. People were screaming and fainting, their effluence amplified through the mountain range. 'You'll feel a warm glow, a heat passing through your entire body,' the preacher assured housewives, seamstresses, grandmothers, librarians sitting on damp soil, waiting for the heat of the cure. Rebekah was sponging it up. Paul kept his eyes closed most of the time. I couldn't tell if he was taking any of this in.

I saw Howells there. I might have said hello if he wasn't damn well down on his hands and knees barking like a dog. His limbs jerked to the tympani of the devil inside his head, as the ministry team held epidermic smiles and open hands above his prostrate body, as if warming themselves on a fire. They made their chant of 'Satan, come out of him in the name of Jesus!' sound like, well . . . *boring*.

Nothing fazed the evangelist, not even an exorcism. He stood on a granite plateau surveying the scene with hands anchored in trouser pockets, grey hair and grey beard framing his face like a wreath of smoke. When Howells began to mutilate himself, banging his forehead on the hard ground, I shared a long doubtful look with Paul. And I gave the ministry team some static. 'What the guy needs is a doctor! A medical doctor.' But they just kept on smiling.

The singing became a dirge. Raised arms polluted the sweet air with armpit sweat. I was ready to leave, but left it one moment too late.

Into that effervescent atmosphere emerged the prophet, Matthew Jones, introduced by the preacher as a new breed of man; born to Ann Jones from Cardiganshire in 1888 when she was fifty-five years of age. An angel visited her during the critical third month of pregnancy and said, 'Daughter, be of good cheer. The fruit of your womb is male whom I have anointed to preach the Gospel like the apostle Matthew of old. Name him Matthew.'

Matthew Jones said the Lord would speak through

him tonight, that he came bearing messages for certain individuals in the congregation. 'Karen Colwell? Are you here, lady? Do I have news for you! The Lord hath told me your husband is a backslider. But the Lord hath put his name in the book. Soon your husband will return to the fold. Your prayers have been answered.' I looked about for a woman who might be behaving like a Karen Colwell should be behaving, but they were all hysterical. 'Only the faithful will be shown the way out of Hell. The Lord tells me these things. I can feel Him planting the words on my tongue. Yea, the Lord is with us right now, in these mountains, His holy grounds, His tabernacle, where Moses received the Ten Commandments. Where is Rebekah Gravano? Show me where you are, lady. Wave your arms.'

I didn't like the way this was going. I said, 'Let's go.'

'It's too late now.' Paul indicated his mother's rapt expression.

'Then may the Lord help us.' I braced myself.

'I see you, woman!' Matthew Jones shouted, pointing in the wrong direction. 'Rebekah, the Lord hath told me to give you this message. Your family has been divided by the devil. Your family is in moral peril and you know this to be. But the spirit will pour down on you all and you will be renewed. Your prayers have been answered.'

Paul and I moved together as she staggered and fell backwards. Our hands were trembling as we caught her.

'Aaron Lewis? Where are you, Aaron Lewis?'

I buried my face in Rebekah's hair, curved my back towards the earth. Yet still that prevaricator came after me.

'Aaron Lewis, I know who you are. You need the grace of God . . .'

I touched the earth with my hands, neared rock bottom.

I waited for the blows, for the stones to crack my skull. Through a forest of legs Saul P. Howells and I met eye to eye. He looked stunned to see me down there. Paul put a hand under my armpit and raised me to my feet. 'What's got into you?' he asked. I chanced to look over to where Matthew Jones was standing. But he had moved on meanwhile, turning to affairs of state and wider prophecies for the nation, his voice squirrelling up mountain slopes with increased urgency as he foretold of another world war about to break, a war in which every man on earth would be called upon to fight.

I was saved by the war ... Or by Matthew Jones. He knew what he knew. Why else did he stop short? To save me being torn limb from limb?

We walked home through the mountain corridor, Rebekah propped between us, guiding her footsteps along the steep and dark path with a crowd in front and to either side. Back in Sharon an hour later the chip shop lights blazed like a beacon at sea.

'I'm fierce hungry, Paul,' I said.

'We have no money.'

'I do,' I said incautiously. 'My treat.'

'Where did you get money?'

'Don't ask me now.'

'You didn't steal it?'

'No, I did not. I swear on my mother's life.'

Inside the chip shop were gathered refugees from a world gone crazy. There was a man holding a seven-foot wooden crucifix between his arms, shovelling four heaped spoonfuls of sugar into his tea and stirring doggedly for a minute to get the maximum performance out of it. Conversation all around was peppery. Rebekah gripped the edge of the counter, confused by the dolphins on the back wall. Under glass, white chipped cups brimmed with animal dripping. Little bodies of battered fish were heaped together.

Lounging on the marble tables was a gang of teenage boys dressed in threadbare duds and oily sweaters, their necks breaking out in boils. By their soporific sprawl I guessed they'd spent the whole day there, loosening the tops of the salt and vinegar bottles. They looked at us expectantly as we settled on green leather benches, waiting to see if we used the salt and vinegar, to see if their sabotage would succeed.

The teak panelled walls had yellowed from decades of tobacco smoke. A menu on the table, once handsomely mounted between plates of glass, fell out of its oak frame as I picked it up. Near to our heads the stained-glass window was filmed with grease. Carved wooden letters above an oval mirror that once read HALIBUT, HAKE, PLAICE, had been rearranged to spell: I HATE PUBIC HELL AKA.

The proprietor's daughter, a thin virginal presence in a white coat, served us our fish and fries with an order of dripping-toast on the side. The boys fooled around with her, flicking flaring matches at her bare legs. Rebekah cut her toast in small squares, rattling the whole table. Paul separated the bright yellow batter from the translucent body of sea cod and nibbled from the edges of his plate. I made even less progress with the chunks of potato saturated in boiled hog fat. Our hunger was great, but our stomachs had shrunk.

Then Howells came in and ordered fries and a sausage to go. A missing shirt button was all the evidence I saw of his recent struggle. He was just real calm dumping the exact change on the counter. He walked out of the shop bolting down the sausage.

Rebekah was staring intently at an empty table. She had

been so silent for so long it raised the alarm in her son. 'Mother, are you all right?'

She pointed to the empty table.

'What? What is it?'

'It's raining in here.'

I listened for rain. What I heard was frying fat, making an agitated sound like a threatened beehive.

'What do you mean, it's raining?'

But Rebekah didn't say anything else. She had descended into one of her darks and never again emerged into light.

I saw the prophet Matthew Jones in the post office the next day as I was walking past. I stepped back to look at him. He looked very small among the stale confectionery, puzzle books and piles of yellowing newspapers, his bald head rising out of a collar two sizes too big. He was nodding profusely at whatever it was Mrs Raleigh was telling him. It was an irrefutable fact that Mrs Raleigh steamed open people's mail, incoming and outgoing. She was a gossip bootlegger, a local Walter Winchell. The post office was her Stork Club, table five-o.

I waited around until he emerged, stealthful and nervous of eye. He gave off the coldness of a frosted beer glass as he brushed by. I gave him a short lead before following him downtown. The street was swept clean of people, not a soul in sight.

Matthew Jones had a motor car parked outside the library, an Austin Swallow with its hood up. Not much change left from three hundred bucks, I'd say. What would a man for whom destiny held fewer surprises than most want from a car like that? I sauntered over as he was cranking the engine.

'Swell car,' I said.

'What?' He snapped upright.

'I said, swell car you've got there.'

'Thank you.'

He turned his back on me, trying to stem my conversation.

'New, huh?'

'Yes.'

'Give you any trouble? Pistons, small ends, crank shaft, that kind of thing.'

'I wouldn't know, I'm sure.'

'Can you hear that noise? Doesn't sound so good.'

'What noise?' He looked nervous suddenly.

'A sort of slapping.'

'Oh, that's just the engine.'

'It's under the sound of the engine, sir. Listen again.'

He listened but gave up quickly trying to hear some note that shouldn't be there. 'I can't hear anything unusual.'

'You'll hear it more clearly when it's driving. If you like I'll take a ride with you.'

'Well, I'm in rather a hurry, actually.'

'Ah, I can hear something else now.'

'Now what?'

'Knock, knock, knock. Very subtle.'

'Isn't that what it's meant to sound like?'

'It's the big ends working loose. If we could go for a drive . . .'

'Oh, all right. Get in.'

He walked around the side of the car, opened the driver's door and laid the chrome handle on the floor at the back, then stepped up into the cockpit. I jumped in beside him. He let the brake off and the car lurched forward. A mile out of Sharon, he said, 'All I can hear is what I think I should be hearing. An engine. I don't think there's anything wrong.'

'Of course it's the engine, but there's a beat in it that shouldn't be there. Like a drummer. You put oil in regular?'

'I had some put in last week.'

'Let's take it a few miles further, then I'll get out.'

The car was all hood and trunk. He manoeuvred it artlessly, unable to hold a straight line. It was a long slow haul rising through the contour lines. Each bend in the road opened on another wall of dripping slate, the clean surfaces contrasting with the craggy fissures higher in the mountains, the gnarly yew trees eking out an existence in the barren soil. Light was failing fast.

'My name is Aaron Lewis.'

'Pleased to meet you.'

'We're already acquainted.'

'I don't think so.'

'Oh sure we are.'

'I'm sorry, but if we've met I've forgotten.'

'Drive on, Mr Preacher. Jog your memory.'

A mile or two further, when we could no longer see one another's eyes, I told him to stop the car. He pulled over and stalled in gear. A river in flood near by, a nightingale, the wind brushing the trees, supplied the overture. 'I fixed General Baden-Powell's motor car once. I don't think I can fix this one. But let's try one more thing, shall we? I'm Aaron Lewis . . . What do you know about me?'

'I told you I can't remember you.' His hands tightened on the wheel.

'Don't hold out on me, Preach. We're all alone on this road.'

'I'm sorry, but if you wouldn't mind . . . I'm late for an important appointment.'

'Mrs Raleigh tell you anything about me? Give you my name for some reason?'

'I am given all names by the Lord,' he scrambled, pressing his back against the door, trying to distance himself from me at the other end of the leather bench seat. 'I . . . I forget some of them.'

'You got an arrangement with Mrs Raleigh? She gives you a few names and the story behind them and you give her money. Is that how it works?'

'Sometimes I need a little help.'

'Ah, now we're getting somewhere.'

'But I can't remember you.'

'You said yesterday you knew who I was.'

'What I meant by that was I knew you were a sinner. We are all sinners,' he added quickly.

'So what makes me so special? Why single me out?'

'The Lord . . .'

'The Lord don't come into this! This is between you and me. I just want you to tell me the truth. Did you pay Mrs Raleigh or someone else for the juice or did you make it up yourself? Or what?'

'The Lord puts the syllables on my tongue . . .'

'Aw, give me a break. Jesus. Get out, get the fuck out of the car. We're going for a walk. Get the fuck out!'

I pushed him out into the road and kept pushing him in the back as we walked. He never looked behind and the only resistance he showed was passing through a field gate where he stopped for a moment. Perhaps the prophet knew more than I knew. I didn't know what I was going to do in the next minute.

We walked down to the river, roaring as it swept over

rocks. I had to raise my voice to be heard. 'Level with me or in you go.'

'I have told you the truth, I can't remember you.'

I pitched him in. He put his hands out to break his fall, dipping his forehead into the water. I waded in after and it didn't feel good at all. Cold and fast. I straddled him between my legs and held his head down so the running surface of the river would shave his nose. I let him feel its cold hungry breath. He didn't make a sound the whole time. 'Last chance to talk . . .'

Upstream, in another lifetime, it seemed, I had enacted the same drama with Edward Manning. Manning had fought like a dog, while Matthew Jones just lay his head on the river's flowing skin like a child put to sleep in its cot. I stared at his old crusty head. Somebody's grandfather, I thought. Older than Sal.

What was there about me now that could be called pure and above reproach? What had gone awry with my mission? I had set off to crack the quarrymen's secrets, discover their tricks of light. There was no light in this.

I yanked him out of the water and sat on the bank with his dripping head in my lap. The prophet saw the change come over me and wept. He raised his left hand and balanced it precariously on my face, gently, as though he were a lover trying to seduce me. He removed his hand and looked at it trembling, as if he'd picked up a disease. 'I do know you,' he said. 'You have an unnatural tendency splitting your life in two.'

Rebekah was committed along with dozens of religious hysterics to Bethesda asylum. Among the inmates trawled in during the revival were *cynfon* turned gibbering wrecks. Thus Rebekah found herself among a community of quarrymen once again. But these were different from the man she had married and the boys she raised. They kept breaking into the female quarters at night to rape the women, a scandal that went uninvestigated, being too dark an issue even to mention in the open.

Paul and I visited her three times each week. The rest of the week we spent in a state of chronic depression. The asylum was a white marbled jug of violent sound, rancid with toxic fluids. The padded wards had slit windows, like the cross-bow orifices in Edward the First's castles. You couldn't even piss through them in emergencies, so most people pissed on the floor. Once in a while an orderly would woosh a bucket of disinfected water over the flags, leaving the inmates to slip and slide on to their butts. Rebekah sat among bodies rotting beneath white gowns, and among women who kept trying to pluck out their own eyes because the world had gotten so ugly and others who rolled back the years until they reached a comfortable time in early childhood to rant about. I wanted to torch the place each time we visited. With Rebekah in it. Paul might have agreed to break her out of there, if the security hadn't been so tight.

We learnt how to shut out these distractions and the stench in order to visit with some kind of dignity, sitting in silence. Rebekah hadn't talked since that moment in the chip shop and neither Paul nor I were able to think of anything neutral enough to say. We passed the hour watching orderlies sitting in their offices like animals in a zoo.

Too much seriousness makes people go mad. They get sent to the asylum and then you lose them altogether. Paul and I kept staring at her, wondering where she'd gone off to. Her face was open for business but the rest of her had left town. The mad can do that, leave their bodies for their kin to lament over. Paul liked to think that she was orbiting the town of heaven. I wanted to press a pillow over her mouth and save her the wait.

The day Rebekah was committed, Sal stopped going to work. He took to his bed full time. Between the two of them, Sal and Paul, they turned the house into a mortuary. It was me who broke the rules by taking a bowl of tomato soup in to him, which he never touched. He just wanted to know how his wife was. 'She's very poorly, Sal, but not to worry,' I said. 'She's okay, compared with how she's been. Top dollar. Yeah, nothing like how bad she was.' But that cut no ice.

He and Rebekah had a symbiotic relationship. As she deteriorated, so did Sal on the other side of town. He grew faint and ate nothing. I took him bits of cheese, like feeding a mouse, only to replace one morsel with another fresh piece. Paul never went in there, which I thought very hard, since the old man was obviously dying. The least Paul could do was put his head round the door.

I shopped downtown with Sal's money and even did some simple cooking. I replaced Rebekah as the woman around the place. I took pride whenever I managed to save a few pennies by selective shopping. And I maintained the kosher kitchen idea: Sal's food one side of the parlour, our meagre supplies on the other side.

One morning I walked into Sal's bedroom with tea and a slice of bread and butter on the wooden tray that Jacob had made as a kid in Sunday school. Sal was on his back, propped up with pillows, his skin a tortoiseshell grey. His eyes were wide open. I lay the tray on the bed and touched his forehead. He was ice cold. I placed my ear under his

nose. Sal had gone during the night. I sat on the edge of
the bed and started eating the bread and butter.

I took Jacob's tray downstairs. Paul didn't move for a
while after I'd told him. He didn't look as if he was even
breathing and for a moment it seemed as if I'd got two
corpses on my hands. He exhaled a lungful of emotion all
over his chest. 'I'll get the doctor in. He needs to confirm
it,' he said and walked out.

While Paul was looking for the doctor I set to wondering
where Sal had gone. I couldn't accept that he was finished
somehow. The heat of the stove died very slowly on my
face. It seemed improbable that Sal was finished up altogether. History was his soul and I thought about that for a
while. History never ends and nor would Sal. Maybe Paul
was incapable of mourning for his father, but I certainly
wasn't. Without shedding any tears (crying does no good
for a man, it's like losing blood, to be avoided at all costs),
I remembered Sal for being the first man kind to me. As
one outsider to another he'd shown me hospitality and the
hand of friendship. Without Sal I could never have carried
out my mission. I wished him peace wherever he was.

Around the same time as Sal passed away, Rebekah had
been pacing about the asylum, peering through slit windows, seeking a way out. Then at a time of night when the
orderlies got banzied on morphine elixirs, Rebekah left by
the tradesmen's entrance, taking with her quite an assortment of women who still had legs to walk on. They plodded
on by moonlight in bare feet until they were eighteen
hundred feet up in the mountains, rounding on the head of
the quarry.

I have been in those mountains at night and know how it
affords a wonderful serenity. With the stars all out, the
scree, gorges and peaks are persuasively comforting. The

mountains at night remind you of how futile struggle is. Mountains don't blink an eye at anything, they remain in a state of complete stability.

I like to dwell on Rebekah's final moments, for they are ones of triumphant peace. I close my eyes and follow her, stretched out with her fellow inmates in one long line, holding hands, before jumping. I see them travel through the dark, a nexus of lost souls in muslin dresses up around their necks, falling through a volume of stars and crashing into the peacock lake below. An instant of self-realization followed by blinding light, by the deepest of blues, impenetrable space and then . . . And then, who knows what?

6 NOVEMBER 1959

'You think I'll like California then?'

'Are you kidding me! Great weather all year round, ocean one side, mountains the others, semi-tropical fauna and flora, surfer's paradise ... Oh man, where does one stop?'

We are sitting hunched over cups of dark brown tea in the chip shop, moments after I proposed that he come back with me to the States. If I'm giving him the hard sell here, it's not because I want a catamite. It's that I wish to see him emerge from the dark ages into the light. I want to do right by him. I can swing a visa, no problem. I can invent a profession of any kind to secure working papers.

'Don't you have someone waiting for you there?'

'You're my main squeeze, Glan.'

'After just a few days?'

'These have not been ordinary days, not in my book, not by anybody's standards.'

'Do you live on the ocean?'

'I live in the Santa Monica mountains. Between Highway 101 and Mulholland Drive. I got me a Negro gardener comes in twice a week to tend my bougainvillaea and my swimming-pool. His wife cleans the house twice a week. She brings their kids along and I let them swim in my pool.'

Like I said, I want to do right by him, which is the absolute truth. But, you know, what the hell, I indulge in a little fantasy: coming home at the end of the day to find

Glanmor in the pool, a hard-on waiting for me as I surface from my dive. Tropical colours, no rain, blue sky all round. And me and Glan making the airbed suffer under our weight.

'Tell me about the Pacific Ocean,' he asks and flushes in anticipation. The mere words *Pacific Ocean* occasion in him an almost delirious awe.

'The waves are magnificent. The tides are negligible. The surf rolls in all day long over the reefs. Dolphins kiss your feet out there. I tell you true. You'll never see a single cloud blemish the sky. Surfers sleep in wrecked Oldsmobiles on the side of the road under Bill Durham chewing-tobacco signboards. Just to get the early morning surf. Some live in caravans with no roofs. Walls is all you need. Like I say, it never rains. Or they camp in the foothills of Mesa Peak alongside religion freaks. I've seen them out there. They live off fruit from orange trees. From olive, lemon, palm, eucalyptus trees, growing wild. Nothing bears down on these guys; they just hit the water. Sleeping in an assortment of smashed palaces, they might *seem* destitute, you know, but they don't feel that way.'

'Amazing!' Glanmor is thrilled. 'That is dedication.'

'Yep, they give up everything to do it all the time. These aren't poor boys either, but sons of oil tycoons, heads of film studios, aerospace industry. They just walk away from it all. For their cashflow needs they plant one of their own people in some diner. Then all they have to do is go in and lay out a quarter for a hamburger and get a buck back in change hidden under a scoop of pecan ice-cream.'

'That's amazing,' he repeats.

'If you're coming, we'd better go soon. The immigrants are taking it over. Come from Mexico or Puerto Rico, work for shit, live in South Central or Watts. All with the same hopes. Few get rich. Something's going to give. The Negroes and the Chicanos are going to outnumber us one day. California's gonna sink through the floor with all the extra weight.'

Glanmor gets ever more daring in public places. Since I invited him to America he's acquired a devil-may-care attitude. We go see a movie in Jerusalem chapel hall, but are only marginally engaged in the film, embroiled instead in our own cock-drama. The air is pungent with the sweet smells we make. Around us some old strike veterans have come just to sit in the dark, and some single women. Halfway through the Elvis Presley film two baptist deacons invade the theatre and start reading aloud from the Book of Revelations by candlelight. They catch men and women, strangers to one other, sitting in too close proximity in the dark. They look at us briefly ... but we're okay, beyond suspicion, and they leave us be. What should occur to them just doesn't occur to them.

We leave anyway. Outside I stop to buy a local newspaper from a young boy, drawn by the front page headline: INCIDENT RATE OF VENEREAL DISEASE IN COUNTY HIGHER THAN IN ANY OTHER COUNTY IN ENGLAND OR WALES – OFFICIAL. The story unfolds on page three. Deacons and their congregations have been conducting a vigil outside the hospital, the 'special' clinic, and printed in bold in the lefthand column are the names of men and women seen leaving, wearing 'countenances of guilt'. People on street corners are checking their copies to see if their names have appeared. The incident rate for the county was 18 in 10,000 compared with 12 in 10,000 for the rest of England and Wales. The editorial interprets it this way: *We live in an age of terror that threatens to strike the future of the human race. We must not rest until those eighteen sinners have been hounded out of the county.*

'Hellenist sex is safer, you know. Queers don't catch the pox. Does your old man know, by the way?'

'No, of course not. You must be joking.'

'Why? He's not going to turn you in to the law, is he?'

'I don't know. He might.'

'Have you tried telling him?'

'Why are you so eager that I should tell him?'

'Let me explain something to you, Glan. Your father was once my great friend. He was never anything but good to me and his hospitality and trust I abused. I even came close to killing him once. He never knew any of this. I regret it now, of course. But I can't turn the clock back. Truth cuts a wound, but lies can kill. Kill a community. Take it from me, I know what I say.'

'Did you tell *your* father?'

'I told my mother.'

'I don't have a mother to tell.'

'My mother said it would bring me nothing but misery. But from then on we had a very special relationship. Maybe in time you could have a very special relationship with Paul.'

Glanmor and I face Paul across the table. Glanmor reads to him by the light of the fire, from the Book of Genesis. The radio plays softly in the background. Brahms or Handel. I get the talking going. I never stay silent for long. Silence has always hidden many dangers for me.

'What's happening to Elusen's grand house now?' I ask.

'The National Trust took it over,' Paul replies.

'As a monument to the slate industry?'

'A monument to the English aristocracy.'

'I hope they still got his enamelled billiard table,' I say and realize too late what I've said. 'I'd like to see that,' I add quickly.

I wait a moment. Nothing happens. Nothing changes. I think I'm in the clear and close in on private celebration when Paul slowly revolves his head, like he's heard someone falling off the quarry face. His eyes widen and his presence unfolds to fill the corners of the room. Paul holds his blind smile on me for an uncomfortable length of time. Behind that smile he goes to work, examining my words for faults, hairline fractures. My own eyes drop like lead weights. He uncoils lethally a moment later.

'How do you know about that?' he asks simply.

'About what?'

'That his billiard table was enamelled?'

I laugh a second. 'Why, is it a secret?'

'Everyone knows what kind of billiard table he had,' Glanmor intervenes.

'No one saw his billiard table until the National Trust opened the house to the public. That was last year.' Paul speaks with a hoarse whisper as though the revelation has winded him on its way through his body. 'If you had paid attention to the details of history, Glanmor, you would not be the fool you are today.'

Glanmor stands up and his chair falls backwards. He is blazing. He isn't going to let his father humiliate him in front of me. 'I've found out why you don't like me. Aaron has taught me something about myself. Which offends you. About my nature.'

'What do you know about nature?'

'There is no place for me in your Scripture-driven world.' Glanmor taps his father's Bible. 'From the very first Book I am condemned.'

We face off for what seems a very long time. I hear myself swallowing. All that has been implied hangs like live wires in the dark electric of the room. Then Paul's immense presence shrinks. He turns to stare into a distant past looking for an exit through which he could squeeze his considerable burden. His eyes float like small boats in a choppy sea. His hands are crossed on his lap. He shivers a little. I watch him bleed an unbearable sadness, his eyes rimmed red. He, who has loosed upon the world so many millions of portentous words is derailed by a few simple ones. They bury him under their weight.

What a fool I am, thinking truth should be a component of friendship. Truth is so often pernicious. Maybe it's a desirable correspondence between politicians and their con-stituents, but it destabilizes friendship. Only moral children and cruel lovers tell cruel truths.

Our irruption in the guest-house is the most violent yet. I keep a two-hand clutch on the windowsill as Glanmor chews up my ass, and a steady fix on the starlit mountains falling away to the west. Where Paul used to lord it. Neither of us says anything, but Paul is the focus for each of us. We are both trying to get closer to him.

Like other men in Sharon, all Paul ever wanted was to grow old in a predictable way, see his son imitate his life to the letter, to the last detail, in work, play, whatever. He wanted to see his progeny walk down that same mountain path from the quarry at sundown, sit in the same pew in chapel, listen to the same sermon, drink the same pint from his personal tankard in the Blue Boar. That's fine. I don't have a problem with that, but it has never been my ambition. Only men with strong religious conviction can suffer such a repetition of routine. But it's no good for the rest of us, the agnostics who fear death, the ultimate failure. That's why I've chosen to live in different societies. If I had children of my own I'd make sure they never repeated my life, however good. I don't even drive the same way home each night. When I go to the stores I walk back to the car on the opposite sidewalk. Every day should be a new day. Nothing should remind you of yesterday, when we were all a day younger.

7 NOVEMBER 1959

At this height of three thousand feet, reached after a pre-dawn's climb, way above the quarry head, with the whole mountain range rolling away beneath us for miles and miles, in an infinite number of surfaces and tones, I am reminded of a story about an American painter named Thomas Moran. Moran was invited to illustrate an article in *Scribner's Monthly* called 'The Wonders of Yellowstone' in 1870. He'd never been to the Grand Canyon and what he produced from someone else's recollections was just flat, dull. Three years later he made a trip out there. The painting he made from the sketches on that trip . . . Well, you know, the wonders were all there for others to see. Except he wasn't satisfied with the painting. He didn't think he knew the place at all. For eleven years he kept returning to Yellowstone Park to make studies of rocks, colour, light, water, fauna and the interplay between them. The visual complexity of the place never failed to overwhelm him.

We drop down a thousand feet before the sun is due to rise and stand at the Gravanos' bargain on the Fitzroy level of Sevastopol gallery. Faint images of Paul, Sal, Jacob ghost the slate. It is a haunted place, a treasure trove of still lives and distant voices.

Pinched between Glanmor's feet is an old paint tin. He prises open the lid with the edge of a penny. There is clear sassafras oil inside. Glanmor immerses a hand and smears it over a few square feet of rock. He strikes a match against the wall to light a large white candle, holding it upside down until the wick is well aflame, then moves the candle two feet away from the oiled rock and spins me round to face the reflection in the wall. With a powerful grip on the back of my collar he moves me from side to side in order that I may view the light pattern, not head on but from the side. The ray of yellow candlelight partly reflects at the surface of the oil and partly refracts through the oil. The refracted light reflects at the slate surface, back up through the oil. These multiple reflections/refractions produce a pattern of coloured light, a spectrum. Glanmor tells me to move my head until I can see just the one colour – a green or a blue. He explains that the longer the reach and the deeper the colour, the flatter is the surface of slate in that particular spot. I focus on a rich blue seam of light, somewhat like the blue of stained-glass windows in medieval English cathedrals that has never been duplicated. After about an hour of this, Glanmor throws the candle into the can. The flame extinguishes in the oil. Then he kicks over the can. It will never be used again. Oil pours out and sinks through the ground.

This is the secret of the quarrymen Elusen had been so eager to discover. In centuries past, monarchs, politicians, used optical illusions as an instrument of government. They kept their subjects fearful by tricking them with interference patterns and claiming supernatural powers. The quarrymen had protected themselves from generations of the Elusen family in a similar way. Listening for a voice in the rock, seeing invisible faults, insisting that a melancholy disposition was necessary to judge the slate, were a smokescreen. Their art had been a science all along.

We return to Sharon with the sun at our backs, pitching

our shadows ahead. I race Glanmor the last stretch home and we burst through the door at the same time in a fit of laughter, instantly swept away on sighting a woman rising from a chair in the parlour, a navy-blue cloak over her white uniform. Paul has suffered some kind of accident in our absence. She speaks so softly I have to get her to repeat herself before I'm confident that I've heard correctly. Paul has suffered a brain haemorrhage. A young boy flying a kite found him collapsed on the edge of the snowline. He tied the string of his airborne kite around Paul's neck and ran to get help. An hour later he was leading two orderlies from the hospital to where the kite hung in the sky.

I cannot for the life of me think straight.

I cannot strike an image out of my mind, of a taut kite-string soaking up blood along its entire length. A diamond of canvas floats high above in the thermals, dripping Paul's blood on to the snow-capped peaks.

The air was full of rain, the wind brittle, indecisive. Clouds broke apart revealing glimpses of the mountain peaks, congealed, broke apart again like dancing partners. The number of pickets had dwindled to a few dozen. Sons of Sharon had left for good. Mothers and fathers were quarantined in their own dead souls. The catcalls, the heckling lulled into bricklike sombreness as infantrymen escorted Lord Elusen and Edward Manning into the quarry. After all they had suffered, the strikers were still overawed by Elusen, his omniscience. From behind a protective line of soldiers he addressed the *cynfon*, now eight-hundred strong, and usurped the moral ground. 'For a long time after the good Lord saw fit to take my first wife, I found a substitute family in my workforce. Now my prodigals are returning to the fold I am a proud father once again. Each day sees more men return to my quarry and to each of you I grant an amnesty. In return you must forget your animosities. Forgive the words and deeds rained against each other. Let bygones be bygones. Make poetry and music like of old. Let your songs soar above the grovelling jealousies of ordinary life . . . let music promote harmony between man and man.'

The infantrymen were limp with boredom, their hands draped across rifle butts, twitching every time a civilian got too close. Some were drunk and threatened pickets standing too close to the gates. Those slouching, braying men from another country disconcerted the children accompanying their fathers, the firearms and lifestyle inseparable to their eyes.

Paul followed Elusen's speech with one in counterpoint. They had never come face to face before. Until today Elusen would not have known who Paul Gravano was. Paul was a now marked man. If the men lost the strike he would never work again. No single quarryman could go the distance with Elusen. He could crush any individual who opposed him.

Elusen listened attentively to Paul from the other side of the gates. 'This strike now threatens our whole lives. It is a trial, poured like the flood upon Jonah's head, testing the very tenets of our Christianity. We shall be better men when it is over, providing we maintain dignity. The path to the promised land leads through desert. Sorrow and despair, thirst and hunger. Which is still preferable to returning to Egypt.'

The usual roar of complicity that accompanied his speeches was missing. There followed a long silence. Men shifted around, showing clear signs of impatience. Saul P. Howells pulled open his neck-tie and cleared his throat. 'It would be nice to go home to . . . This is more of a punishment for my new wife, Paul. That's the very bad part. It's wrong now. I've got a bit of doubt in my mind, whether we should go on. For the first time I've got a bit of doubt. Because my wife's expecting, maybe.'

Paul laid his hand on my shoulder. Brothers in arms. It was witnessed by Lord Elusen, the accountant Edward Manning, numerous soldiers; in front of whom I felt a sharp heat draw up from inside my loins, through my belly and lodge inside my head, fizzling like a fuse. Things were on the brink of anarchy out there. It was anarchy in here too.

'Do you not think I understand how a family can be crucified by this strike?' Paul let an artful silence hang. 'I received a letter today from my brother in Pennsylvania. He tells us "Stay out till death". To go back now would be less than our manhood is worth.'

'"Stay out till death"?' Elusen's voice rolled over the heads. 'There is no famine in Pennsylvania. Follow your own instincts, not paper commands from America.'

Paul was pitched into gloom. In the stand-off between the two men Paul had conceded a round. I thought that maybe now he would move from his fixed position, that tough ethical line. For men are not expected to act like gods. Human nature needs to be frail, needs to be weak, must give in to temptation, otherwise it cannot be called human at all. Paul's idealism, his moral virginity, threatened me, threatened us all. He was a model no man could live up to. He belonged to a community of one. In his ideal world there could be no sickness of the soul. I would have no place there with my incurable disease. He would always be leaving me behind at the pearly gates. Never the twain shall meet. The truth was, I wanted him sullied.

Now we had the house to ourselves I played wife while he worked up to his big responsibilities, devouring books and composing speeches. He did all the important work, changing the world, while I did the cooking and the cleaning. Or so it would have appeared to anyone looking through the window. But like any marriage, things were not what they seemed.

He sat on the edge of the table with his legs apart, leaving me breathless, swimming up into the opening. 'The committee have called a ballot,' he was saying, his voice bereft of its former rich optimism. 'I think they will vote to return to work. After all this, after all we have gone through. I cannot bear it, Aaron.'

'Let the ballot decide now.' I kept my voice down below the shading of desire. 'You have done all you can.'

'No, it's all over. I can smell the smoke in the air.'

For the first time he sounded like a man on the cusp of changing sides; of losing faith and gaining different values – the values of *cynfon*. Once that slide started, anything was possible. He might even roll into me, lying on the devil's side of the bed. The suspense was physically painful.

Paul had grown quite weak in the last few months. His waistline, once a robust thirty-four inches, had lost ten inches. Flesh hung from his cheekbones. His beard was wiry and turning white. His eyes glistening with tears, his trembling lips, the restless hands, the way he pressed down on the back of a knife to cut a lump of soft cake, sent shock waves over me until I could take no more of it. I gripped his wrist and removed the knife from his hand. He withdrew his hand from inside mine. He took his coat off the door and walked out of the house.

The ballot was held in Jerusalem chapel. The pews filled with slow bodies. A pre-ballot debate disintegrated into accusation and counter-accusation until a vote whether the debate should continue had to be called. When that was carried in favour, the men began vying with one another, speaking out of turn, vociferously fighting for airtime. Paul made a last attempt to convince them of his way. 'If we return from the wilderness now into the land of captivity we will have to stay there for generations to come. Remember that we are fighting not just for ourselves but for our children and their children. The faces of the tribes are turned towards the land that is seen, and we will travel on to Jordan, even if we fail to reach the Promised Land.'

The vote was counted and passed narrowly in favour of continuing the strike. It went 461 to 320, with 90 abstaining. Paul had restored his composure in sight of principled aims. I had lost him again. All the more disappointing when I thought I had won.

Leaning against a wall of spruce fir, disturbed by the blackness of trees, Lord Elusen's gelding stamped and jerked its head, snagging its bridle on the needle-like branches. The rain fell steadily, whispering on its way through the knitted foliage. 'If it *is* a test of faith,' Elusen said, 'then I cannot win. I should give way.'

The power to determine history either way dropped into my hands. If I told Elusen to give way, Paul would go beyond all reach. If I told him to hold out, Paul would be plunged into cynicism. He would enter the same dark kingdom that I inhabited. This is how the world sometimes revolves, around two individuals in a drenched forest.

For the moment I fell back on the rulebook: Be loyal to the client; be loyal to America; act patriotically; defend the national interests; resist communism's charms; resist decadence; don't venture into lowly pornographic areas; don't discuss internal affairs; don't lose composure; don't point at people; don't laugh too easily; walk, never run; sit up straight, never squat, don't slouch against walls; don't spread your legs; if you need to spit, spit into a handkerchief; don't trim your nails, pick your teeth, clean your ears in public; don't take off your shoes; never whistle; don't look at your watch; don't talk of illness or anything depressing; don't foam at the mouth . . . But these rules that had protected me against most things, particularly against myself, failed me now.

The power invested in me felt weightless. A dime in the palm to buy candy. In the end I made my choice between a sense of professionalism and a moral quagmire. 'You must stay on the strength,' I said. 'All the talk of staying out till death is rhetoric. The voting is running contrary to the rhetoric.'

'I have lost so much money over this affair. I cannot survive another month.'

'The strike will collapse in a matter of days. I am confident of it.'

'Can you guarantee it?'

'The strike will fold in days. A week at the most.'

Elusen gave me a money bonus. I intended spending it promiscuously. When a man comes by money in such a fashion, it is prudent to spend it as fast as possible. Otherwise he will live to regret what he did to earn it.

The annual grant from the General League of Labour Unions, the last source of funding, was suspended. Families that moved to the USA, the Ukraine, to the South Wales and Yorkshire coal fields, were forgotten. Elusen's estate agent made a count of houses left empty and disclosed a quarter of the stock vacant. Cases of tuberculosis, rheumatic fever, scurvy and anaemia had reached epidemic proportions. Young couples suspended their desire for having families; the strike was an effective contraceptive. Those who could not control themselves postponed the naming of babies for twelve months. Children would earn their names by surviving to their first birthday. Ever-increasing numbers of strikers wrote to the manager, Edward Manning, asking for work, sacrificing a principle for the sake of another:

their purification traded for the lives of their children. They swapped one hell for another.

The strike committee called a second extraordinary meeting. Paul made his pitch. 'If my minister, Mr Parry, orders me to stay on strike, I will do so. I will do so because the minister represents God's will on earth. For me to act independently on something I believe is right or wrong is worthless in the eyes of our Lord.'

Mr Parry replaced him at the lectern. 'I cannot accept that responsibility any longer.' He sat down in a vacant deacon's chair, demoting himself and crossed his hands on his lap.

Paul returned to the lectern. 'As we speak a crowd of *cynfon* are gathered in the grounds of Elusen's home attending a tournament, to see soldiers wrestle on bare-backed horses, compete in a tent-pegging contest and fight mock battles, watching our enemies play the clown. It is a foreshadowing of the new culture about to dawn.' Then Paul sat down next to Mr Parry.

Parry, confused as to what Paul was implying, as indeed I was, approached the lectern. 'I advised you against violence and there was violence. I told you that the innocent would not starve and they starved. I have prayed for guidance and received none. It has gone beyond the power of my judgement, as man and as God's holy witness, to tell you what is the right course of action.'

This time he did not sit in the deacon's chair but dragged his feet through the blistering silence and sat at the very back of the chapel, nearest the door.

The voting went in favour of giving up. I went over to where Paul sat isolated in the deacon's chair, but he would not look at me.

I walked out of the chapel and into a ringing light. The town had fallen after almost two years. My job was done. And not a moment too soon.

It is a long dimmed walk to Elusen Memorial Hospital. I can still remember the smell its incinerator used to make, of roasting marrow and bone. Like the smell of burnt coffee. Principally an amputation centre for a hundred years, it was handed over to the National Health Service in 1948. The chimney has since been knocked down.

A doctor briefs us on Paul's condition in the echoing corridor. 'He's suffered a massive rupture of a blood vessel.'

I'd have expected nothing less. For Paul. I mean, he's a strong man. When something like this was going to happen, go wrong, it had to go wrong in a big way. That's not what I mean at all . . . Jesus. I feel Glan's eyes on me, bleeding me dry.

'It's very serious,' the doctor continues. 'I must warn you to expect the worst.'

Glanmor cannot hold his head up any longer, his body pushes down into the ground, as if the air above him was heavy. He thinks it is all his fault. First his mother, now his father. His revelation ruptured that blood vessel, he believes, while I blame my own revelation.

We find Paul in a ward of eighteen geriatric men. They cough, sleep or stare at stark whitewashed walls. Paul's bed is next to the nurses' station, near to the door. A saline drip feeds his arm, a catheter is stuck between his legs. So helpless. So changed. Those hands that assisted in moving half a mountain rest on his stomach. I lower myself down on the edge of a wooden chair beside his bed and clasp his hands. Glanmor, sitting on the other side, holds his own head.

There is nothing in the world I want more than for him to wake up, even for a second, even if it is going to be for the last time, so I may ask his forgiveness. My face close to his, I hear a faint sound coming from his mouth, like the sound in a seashell. I whisper into his ear, back at the sea. If he can't send out dispatches, then maybe he can receive mine. I call him my brother, my father, my mother, in a confusion of motives. His face is bloated and a startling silvery colour, like a beached trout. He breathes irregularly, barking and snoring alternately. We sit with him in silence for three hours. It is a time of great loneliness.

Going out of the hospital to eat is a picnic of anxiousness. Glan and I walk around outside, lunching on wind. I choke on the breezes while Glan turns his fear on me: why have I seen fit to wear a worsted cashmere suit and maroon silk tie this day? 'What are you trying to do?' he snaps. 'Impress the sheep up here?'

We return to the hospital to find Leah shouldering the icy corridor wall, smoking a cigarette. She has just come from the train station. She weeps as she sees me, has herself a good downhome cry. She has noticeably aged. There is a thickening of the hips and her face is fuller, with lines etched across her forehead and around her mouth. Her bare feet are strapped in sandals and her ankles show a speckling of veins. Her teeth don't look so white either. Then she embraces me and I feel all the years fall away like so much dust.

She puts an arm on Glan's shoulders. Without fully realizing what the issues are, she says, 'You never realize what this is going to feel like, do you? It's unimaginable. Death is what happens to other families.'

Glanmor walks off. He doesn't want to hear this, she's saying all the wrong things, platitudes a stranger would offer.

'He's not dead yet,' I round on her, stamping out the cigarette she throws on the floor.

In the ward Leah wanders around talking to all the other patients, unable to accept the situation. Glan and I hold Paul's hands and meditate on his fallen face. I begin to consider what his life will be like if he does survive. He is already blind. If he lives he will most likely be paralysed. He might never walk again, which would be a living death. Movement is Paul's only sanctuary.

The nurses are all heroic and come in various shapes and sizes, running from patient to patient, while consultants stroll from vending machines back to their offices. They dress like lawyers, like undertakers. It's the nurses who keep Paul alive, they who check his chart every couple of hours, offering kind words to us: He's comfortable; Not much else we can do; It's a long wait; Would you like a cup of tea?

We are at his bedside when Paul's mouth opens and words crystallize on his tongue. 'I can see,' he whispers.

'I don't think he knows what he's saying,' Leah shrugs.

'Yes, I do,' we hear from the bed. 'I can see all. Glanmor, why are you crying?'

'Dad, I'm sorry about what I said to you. What I told you. I want to . . .' He rushes at everything he has been rehearsing to say.

Paul squeezes his hand and silences him. 'Get me something to drink, there's a good boy.' *There's a good boy.* Glanmor is beatified by a cluster of words on his father's tongue, words that he has waited a lifetime to hear.

'I'll get you one,' Glan and I blurt out at the same time, competing against one another.

'Do you want a cup of tea, Paul?' Leah asks.

We could all do with something to drink. Dynamite a lot stronger than tea.

'Who are you?' he says in Leah's direction, wiping the smile from her face. 'The nurse?'

The cruel truth is he can't see at all. I hold my tie in front of his face. 'Can you see this?'

'Yes.' He rolls it between his fingers. 'It's a tie.'

'What colour is it?'

'Red.'

'So why can't you see Leah?'

Leah stands up from her chair. Paul points at her the second she reaches the foot of the bed. 'I *can* see her.'

Paul's vision is only partially recovered. He sees nothing to his left. And so we all cluster on the opposite side, sharing the same passion to be identified, as if we don't really exist unless Paul sees us.

'Where is this place?'

'It's the Elusen Memorial.'

'Get me out of here! The doctors are on Elusen's payroll.'

'Oh my God.'

'Mmm?'

I wander off to clear my head. A nurse attending patients at the far end of the ward I entice away with an account of miracles. 'Paul can see. He wants tea to drink.'

'He's come round? That's marvellous.'

'No, he can see again. He was blind now he can see. Please can you find him a drink.'

She goes into the kitchen without commenting, without even a spring in her step, as though what I had just said was no big deal. Miracles are quotidian in her hospital.

'I want to lie down,' Paul is saying as I return.

'You are lying down.'

'Really? I thought I was sitting in the chair.'

His right hand plays with the left one, holding it up, letting it drop, rubbing the dead arm. He grows more agitated minute by minute, cupping his right hand around his lips and shouting: 'Nurse! Nurse! Nurse!'

'What do you want, Dad? What do you want?'

'I want a nurse to lie me down.'

'You are lying down.'

'Nurse! Help! Nurse . . . I want a nurse to tell me I'm lying down.'

The matron who arrives carries so much weight that she is breathless from the short walk from the other ward. She places a white cup and saucer on Paul's table and sits on the bed to catch her breath. A mass of downy hair on her face is damp with sweat. 'Now then, what's the matter?'

'I can see everything. I can see you. I can see people in beds over there.'

'That's marvellous.'

'Has this ever happened before, Matron? A blind man regaining his sight.'

'He has cataracts, doesn't he?'

'Yes.'

'A haemorrhage can push things back into shape sometimes. It might have released the pressure from his eyes. Here now, Mr Gravano, here's a nice cup of tea for you.'

She raises Paul's shoulders up on to a stack of white pillows and tries to persuade him to drink. Paul takes a sip of sweet tea, swallows and coughs violently. He tries again with the same result.

'You must drink,' matron says, 'or you'll get urinary infections.'

'Leave it with me,' Leah says, 'and go to see your other patients.'

'You drink some,' Paul whispers conspiratorially to Leah when matron is out of earshot.

'It's for you.'

'I'll drink more if there is less in the cup,' he says with ill logic.

His agitation increases to a frenzy and we have to leave the ward. It is just too upsetting. We go and sit in the waiting-room. A man of Paul's age sits near by in his blue striped pyjamas and slippers. Like Leah, he chain smokes. He tells us that he has cancer of the lung. The doctor has given him a month to live. He does not sound like a man should sound who can foresee his own funeral, is not overly subdued telling us the news. He could be out now pillaging but instead talks about dying, as if it was an unexpected change in climate, as if life is so morbidly empty that death seems a continuity.

I feel trapped in this place where all the talk is about death. I feel claustrophobic as if all the world were dying. And where there is death there is guilt.

The matron rounds us up an hour later, bearing brittle news. 'I'm afraid he's had another little bleed.'

We all leave our seats at the same time, mouths open.

'This is very common I'm afraid. There is nothing we can do.'

Back in the ward, sitting it out beside the bed. Glanmor and Leah hold his one sensate hand as I talk to him in his sleep. It worked the last time. If only he would wake to tell us that he wants the nurse to tell him what side he's sleeping on; just that. Is that asking so much? Of Paul?

The bleeding must have stopped as his eyes open. They search for a way out. It is midnight exactly. He opens his mouth and asks for toast. I don't think I've ever felt such rapture. Glanmor hugs him around the neck. I run through the darkened ward, calling out for bread.

Paul gets his wish a half hour later, by which time he has become distressed and confused again. He tries to speak but the effort is so exhausting that he falls asleep before he's completed a sentence. This pattern keeps repeating itself. He wakes and calls and falls asleep. For a few seconds each time. He seems on track for another haemorrhage. A moment ago we wanted him to talk. Now, just as desperately, we want him to sleep. We sleep in the hospital, fitfully, in hard wooden chairs beside his bed.

The first thing I do in the morning is go out and bring in delicacies to tempt him. Pickled herring and brown bread, jellied eels, homecooked rice pudding. He turns everything down except the rice pudding. This is encouraging. He takes a spoonful and retches, his chest afloat with blood and mucus. Leah moans and Glanmor excavates with his fingers morsels of rice from the corner of Paul's mouth.

I have a pack of cards which I withdraw slowly from their box, unsure whether he will approve. I shuffle the pack and lay down a game of patience on his bedside table. He watches me with some interest. I start moving the cards around and Paul taps on a king of spades with a fingernail, then on the queen of spades where the king should go. Glanmor and Leah smile through their tears. We see hope in the minutest of advances.

His interest in cards grows stale. He starts thumping the bed, himself. Nurses come and they go. At first they try placating him by turning him over. Then they try to fool him by shaking the bed, then by rustling his sheets, until even they can't take it any more and fly by without stopping on their way to other calls.

Two doctors stroll by around lunchtime and yawn at the

foot of his bed, ignoring him, ignoring us, as they discuss his chart like a shipping forecast. 'Would you take some notice of him, please? My father ... This man here,' Glanmor starts. 'You think he's just an old broken body taking up space in your hospital.'

'We are doing all we can.'

'Doing what! I'll take him home if this is all you can do.'

'That would be against our advice.'

'What is your advice?'

Paul overhears this discourse, his eyes widening in terror. 'Are you doctors on his lordship's payroll?' The doctors look on, beyond Paul, to their lunchbreak. 'Go away. Just go away. You'll tell Elusen I'm too sick to work. I know. I know how it works.' Paul pokes Glanmor hard in the arm. 'They'll be bringing the dinner around soon. But I'm not going to eat it. They make me eat that rubbish when you've gone. Don't let them make me eat it.'

None of us ever considered that he could be reduced to this. We cannot square what we knew of him with what he has become in just a few days. That pair of bulbous frightened eyes in a shrinking face. A body wasting away by the hour beneath a sheet, emptying his bowels in bed. His mind a room of moths, of partially remembered things. On his table Leah arranges a vase of flowers. Beside the vase is his watch, a beaker of tap water, a blood-stained handkerchief. How did his great life become so diminished?

Two nurses draw the curtains around the opposite bed after the patient there dies. 'For want of breath,' Paul says, leaning out, catching on. 'That's how your mother died, Glanmor. For want of breath. What time is it?'

'It's five o'clock, Dad.'

'Another hour till sundown. What time is it?'

'I just told you.'

'No, you didn't.'

'It's five o'clock.'

'You still here, Leah, when any decent woman your age would be in bed?'

THE STRIKERS DEMONSTRATED HEROISM AND RESTRAINT FOR THE PRINCIPLE OF THE LOYALTY OF ONE QUARRY-MAN TO THE REST – *Manchester Guardian*, 4 September 1939

QUARRYMEN HAVE WASTED THE HAPPINESS AND COM-FORT OF A CONSIDERABLE NUMBER OF YEARS OF THEIR LIVES IN A FUTILE EFFORT TO OBTAIN TERMS WHICH THEIR COMMON SENSE MIGHT HAVE TOLD THEM WAS OUT OF THE QUESTION – *The Times*, 3 September 1939

I sit in the local library reading old newspapers. A single paragraph in each paper, lost in the middle pages, was all the strike merited. Events in Europe eclipsed the drama here. War broke out four weeks later and within a year slate was declared a non-essential industry. The quarry became operative again, on a limited basis, after 1945, when the hand chisel was replaced by the hydraulic drill and a new breed of quarrymen replaced the old, decoding their myths with geology and driving to work in private cars from villages fifteen miles away. But Elusen had lost his world monopoly by this time. Competitors from France, Spain, Canada, the USA entered the field during the strike and took away his markets. In 1948 it closed its gates for good.

Another article, in the *Genissen*, catches my eye. Between 1820 and 1935 the only 'serious' crimes recorded in Sharon and district were an attack by a man called Owen Rogers upon his wife and a bicycle theft twenty-seven years later. Sharon was not like other towns tossed up in the exigencies of industrial expansion; it was not a gold rush centre. It grew out of a rural society. One language, one industry, one religion and no dissenters. No pickpockets, burglars, cockfights, dogfights, pugilists. Three generations of slate craftsmen had created an obedient society. No civilization can afford to destroy communities like these, lose their example. But that is exactly what happened. This land they once called the Land of the White Gloves is now the devil's pride and joy. Two millennia in the making, two years to destroy.

Sunlight seeps through the stone-framed window in the library. It illuminates the mildly obscene heterosexual graffiti carved into my table, and falls across the yellowing newspapers, crumbling in my hands. I feel the warmth of the sun on my face. I close the papers when Glanmor arrives. We tiptoe past the bespectacled librarian snoozing behind her desk and sneak a frightened kiss under her nose. We walk out into the cold sun and head back to the hospital.

Paul traces his finger across the pages of the library books I brought in, but the words fail to come to life, are brush strokes of an abstract painter. He looks at me mournfully. 'The devil slipped ants under babies' tongues and cast a red shadow. The babies woke up in a hornets' nest.'

The young minister from Jerusalem chapel comes to visit Paul. Not old Mr Parry but his successor, a round-faced thirty-year-old with a neatly trimmed beard and a vacuous accent. He is full of physical hubris. I am in the middle of shaving Paul when he arrives.

'How's the patient?'

'If you stand any further back from the bed,' says Paul in a moment of alacrity, 'you'll be out in the street.'

The minister creeps forward. 'Hello? You're looking well.'

'I'm feeling tired. I want to go to sleep.'

'Of course you do. You must be very tired, eh?'

The minister says a prayer and disappears out of the hospital.

'What was all that about?' Paul asks.

'He was saying a prayer for you.'

'A what?'

'A prayer.'

'What's a prayer?'

On the fourth day Leah and Glanmor decide to discharge Paul after a succession of secondary illnesses – chest cold, urinary infection, dysentery – drag him closer to the end. He just can't hit a clear spell in this hospital. Like the priests, the doctors fail him. They make too weak a case for keeping him. He's going to die, they all say, one by one. A little infection could finish him off. 'We certainly can't do worse than that, can we?' Glanmor confers with Leah.

We take turns to watch over him, our life dominated for the moment by the culture of illness. Glanmor makes a bed up for him downstairs. He cares for his father as well as any nurse, while Leah has difficulty, as I do, with the toileting detail. It is hard to show unconditional love to anyone in this state. Paul's former character, the recipient of our erotic love (for all love is erotic), has been debased through sickness. Fluids we spend a lifetime concealing, that rightly belong on the inside, erupt from every one of his orifices, even his ears. But if anything, Glanmor, who owes his father far less than I do, loves him more than ever before.

My shift comes round. I sleep only as long as Paul can, which is never more than an hour at a time. The mantel clock ticks on, says it all, as I try to feed him liquids through a straw: soup, milk, water, tea, barely wetting his lips. He spits just about everything up. I make jello as a way of administering fluids, which works for a couple of minutes until he refuses even that. 'I don't want to see a jelly for the rest of my life,' he says. How long is that? I catch myself thinking. The egg and milk custard I make is taken from a recipe written in Rebekah's own hand. 'This is good,' he appreciates, and manages to keep down a mouthful. 'About time you made one of these. Close the curtains . . .'

'They are closed, Paul.'

'Are they? Are they really?'

The curtains lie to the paralysed side, his blind side. Whenever I desire privacy I go and stand by the window. So he'll forget exactly who I am. So I can forget who I am.

Leah comes downstairs, rubbing her swollen eyes and goes straight outside to smoke her first cigarette of the day.

'What's Leah's child called, Aaron?'

'Jake. Not a child any more either.'

I join her on the porch and we both stare at the mountains, where we once walked out together. 'What are you going to do with this house, Leah?' I ask. 'Eventually?'

'My family's paid rent for a century. Sal, and before that, Rebekah's father. Paul bought it from the estate. Now it's ours, but I don't want it.'

She tells me that she has found a reasonable life for herself in Liverpool. Her son is a young man now, sailing in the merchant navy, like his grandfather before him. It was Leah's idea that he go to sea, far away from the certainties of land. No two days at sea are alike.

The minister makes his second appearance since Paul's stroke. He comes to the house in funereal black with a fanatical deacon in support. Leah stiffens like a cat when a dog walks in. She goes and stands at the window, invisible to Paul. The silence in the room is like falling snow. The deacon watches Glanmor run towards the stove to save a pan of milk from boiling over and says, 'What's this then, boy? Role reversal, is it?'

'God bless you, Paul,' says the minister.

'God loves you,' says the deacon, his eyes unfocused, undiscerning.

'God will heal you, Paul,' the minister says.

'Excuse me, minister, but it's me that wipes my father's arse,' Glanmor erupts, his voice trembling.

Paul asks who he is talking to.

'The minister and his grocery clerk.'

'What do they want?'

'They are trying to sell us religion.'

'Salesmen? What kind of job is that for a grown man?'

It has become clear to us that Paul has misplaced his religious belief as a result of this illness. It is another remarkable by-product of the stroke. Brain cells that once contained the words of the Lord have been terminated, like a library shut down. Simultaneously he demonstrates a measure of unconditional affection towards his son for the first time. Paul has replaced tyranny with a new-found tenderness. He has lost religion and, with it, moral indignation. He makes strides in Glanmor's direction, through the minutest of gestures, salvaging the lost years in tiny parcels of words. Through illness they each find salvation.

Paul makes Glanmor laugh as he feeds him water on a spoon. 'If ever we get through this, Glanmor, you promise to take me surfing?'

'All right.'

'You're good at this nursing, Glanmor.'

'I wouldn't mind being a nurse.'

'But don't forget to take care of yourself.'

'I won't.'

'You should go out more. Take a holiday.'

'I might go to America . . .'

Paul raises his head off the pillow and gives Glanmor a surprised but not hostile look. 'That's where my brother ended up! America's a great invention.'

'But not yet.' He glances over at me.

'I want to say something to Aaron,' Paul says. 'Aaron, you came among us at the wrong time.' He shuts his eyes for a second. 'What time is it?'

'It's seven fifteen.'

'Of course, summer time. Down tools!'

All night long he chants 'on my right side' and 'the slate will slice like butter', without respite for nine and a half hours. As a feat of sheer energy that is quite impressive. Each time he calls I rock him back and forth until my back gives out from leaning over the bed.

He has acquired a smoked look in his eyes. This is how I know it has started. He is deteriorating. His soul is leaving through the eyes. I wake Leah and Glanmor and we begin to prepare ourselves. We change his sheets and wash his body, but there remains a smell we cannot remove. A stench of urine. A stench of dysentery. Indelible hopelessness.

Paul holds Glanmor's hand. 'Your mother would be proud of you.'

'Don't. Don't.'

'But she would. She would be very proud of you. You are her boy. She was good at caring too. Can you hear that?'

'What?'

'What's that noise?'

'I've got the wireless on, Dad.'

'No, it's snow falling. That's when your mother and I conceived you ... When the snow was falling silently outside.'

He closes his eyes and for a moment I believe he will recover. Five minutes later he opens them again. 'I've had a dream. A marvellous dream. But I can't remember it. Where do all our dreams go? All that energy has to go somewhere.'

'Perhaps that's what we mean by the afterlife,' Glanmor says. 'A space beyond eternity with no boundaries, where a web of memory exists, the memories and dreams of every man and woman on earth.'

Paul smiles. 'My son. My beautiful son,' and kisses him fully on the mouth. Later, Glanmor tells me the sensation that passed through him was of Paul's dreams and aspirations, siphoned through the contact of their lips.

When Glanmor draws his head away Paul is dead.

In his eulogy the minister mispronounces Paul's name. 'Paul Gravanah,' he says, 'was born in this town and worked in the quarry until the Great Strike. A keen family man, his hobby was walking.' When history is left for preachers to tell, the dead are unjustly served.

There are a few of the old Sharonites sitting in chapel, whose eyes roll in response to the minister's epitaph. Even though it is their show, Leah and Glanmor bow to seniority and sit behind them. The detail that will stay with me longest is of these old men's calloused and speckled hands, sprouting from white shirt cuffs and sleeves of black suits, wrapped around the thick waists of their wives. 'Although we grieve for the passing of this man from the world, it shall cause less suffering to us than to those who believe there is nothing after death. Paul Gravanah has now embarked on the Christian journey. His soul is destined for another place. Amen.'

Leah and Glanmor choose to have him cremated rather than bury him in a graveyard that one day would go the way of the rest of the town. Leah suggests the quarry lake for scattering the ashes, below the golden peaks. The lake will be here long after cemeteries have been cut out of the ground.

We take turns to carry his urn as we climb up the mountain. At four in the morning, Paul's remains are as light as kindling in our arms. His last kindness to us. The sun rises, turning the dew-soaked slate into glass, mirroring our image back to us. We each taste Paul's ashes before scattering them over the surface of the peacock lake. Glanmor hurls stones into the water, trying to sink the congealed ashes. They float for several more hours. We sit, not so much unwilling to leave as we are uncertain of which direction to take next. Just when I think that we might still be here in a week's time, a bank of storm cloud appears in the south east. It rolls over Sharon and comes our way. We begin to hurry down the mountain as the first frozen drops of rain sting our faces.

I DECEMBER 1959

During the past few days America has launched another monkey into space. The monkey returned to the earth's surface in a capsule but is lost in the Atlantic, while several miles below on the ocean floor, US marine scientists in a deep-sea bathyscaphe are searching for new forms of life.

So why do I identify with the monkey?

We walk from the station through Liverpool docks. Through a valley of giant Californian redwoods lying on their side, between mountains of jute sacks and steel cable. Glanmor is still dazed. He has dived to a depth where he cannot be reached and trips over hawsers and tracks. Hulls of ships in berth lean on me – sheer walls of painted steel, stained with orange rust. Rivets vibrate from the engines. They fly the flags of Arab nations, Israel, China, North America, South America, USSR, their names and countries inscribed on transoms reflecting in the still water of the dock. In scale they are like small towns, and lifebuoys tied to the stanchons serve as a reminder that peril is always around. The quayside is chock-full of men steaming from exertion, bumping and scraping and swaggering about, hauling great weights, emptying cargoes into granite-rimmed docks. It's reminiscent of another time, another place. A foreman rests his elbow in a manacling chain hoop embedded in a store-room wall, overseeing an all-Negro gang who sing a six-part harmony while unloading a lighter on to the quay. From warehouses ooze smells of Colombian coffee, Darjeeling tea, perfumed essences. Each time we pass through the different zones of smell, my sense of place gets thrown out of kilter. It is a suspenseful place to be, dramatic. Leah's buoyant rhythmic step keeps time with the beat. Clearly she's found what she's been looking for.

There are so many ships anchored at the Bar waiting for a berth that I lose count. Ever more vessels come into the river, flying their bunting and announcing arrival with hortative blasts on steam whistles. Tomorrow is Friday and sailing day for CPR, the Anchor and Ellerman Lines, Cunard . . . And sailing day for Glanmor and me.

We emerge on the other side of Queen's docks into Upper Pitt Street among the doorstepping Negroes. Into a small cobbled courtyard and Leah holds her smile, but I can't raise one at all. The fact that this place is a slum doesn't bother her, but it sure bothers me. Six houses face each other across the courtyard. The Nigerians, Gambians, Senegalese, who live there, share the one toilet backed up against a brick wall. Glass is missing in more than one window, behind which rabbit skins dry in the flow of air. Her front door is open for one and all. Dominating the parlour is a black upright piano.

My legs start shaking and I dump myself on one of the chairs behind a rickety table. 'So where's the family?' I ask. 'Where's the main man?' She doesn't answer, but initiates us into the rooms upstairs. Above the parlour are two more storeys. To reach the top of the house we first have to climb through her bedroom. Two single beds have been joined together. Clothes and sheet music are strewn across the floor and standing in the window is a sickly geranium in a flowerpot. A tinted photograph of Leah's son, Jake, in deck officer's uniform, stands on the dresser in a silver frame. The resemblance to Rebekah is obvious, in the full lips, long nose and glass jaw. Somewhere not far off is the sound

229

of drums, from more than one source, and this completes
my disorientation. From the window I can see a neighbour
moving around her place, so close I could borrow a cup of
sugar by leaning out of the window. 'You left a home with
a view of the mountains – for this?'

'We're living friendly. It doesn't bother me, frankly.
There is more here to keep a girl lively than in Sharon.'

Leah has forfeited her citizenship in the white kingdom
coming to live here, where she will always be an outsider,
the outsider she was always trying, without success, to
become in her home town. A simple solution, a long time
coming.

At the very top of the house, Glanmor and I put down
our bags. He tests the mattress for softness. Climbing back
downstairs, Leah starts telling us the story of her new life.
She is keen to impress upon us that she's had it hard before
she had it good. I suspend belief. What can be harder than
this?

She came to Liverpool in the beginning to find a place
for herself in the big world where men and women had
some life about them. But she had no skills to make her
own way and was soon destitute, with a baby to care for. It
was like discovering she was bankrupt. Nobody warned her
about that before she left Sharon, she says. I did, but let it
slide.

Then she met this guy called Norman, a low-rent, white
nigger and the big world got reduced to its lowest common
denominator. She and the baby moved in with him in his
studio on Stanhope Street and from then on all she got out
of Liverpool was a bellyful of his spunk every night.

Norman was in the building trade, something to do with
home extensions. He never really explained and whenever
Leah asked he'd get curt. 'I go to a home. I move one room

and put it somewhere else. End of fuckin' story. What do you want me to say?'

Norman was the first person she'd heard swear like that – with an undertone of violence. But whatever his business was, it cost more than it made. When official letters came through the door, Norman threw them unopened out with the trash. Sheriffs . . . the bailiffs kept appearing at the door. Norman would go hide in the bathroom and send her to deal with them with a pillow up her dress to make herself look pregnant. The sheriffs informed her the studio was registered in her name.

Norman told her not to worry, he had a house over in Picton rented out until such time as he could get his business up and running. Then they would move in there together. He took her out to see it when Leah called his bluff. A Georgian house in bad shape, but at least he had the keys. Norman had her running all over the place, showing off the fitted wardrobes, a white-tiled bathroom he'd put in himself. A woman's lingerie scattered everywhere he dismissed as the property of his tenant. Leah asked if his tenant minded the landlord letting himself in when she was out. 'All this will be yours one day,' he said.

Norman was so possessive he wouldn't let her go out anywhere when he was gone. He wouldn't even let her take a stroll down the landing to talk to the neighbours. When she mentioned she'd like to go home to see her brother and nephew, Norman went into a sulk. This was a sure sign of a slip in her total dedication to him, even though she'd invited him along. 'Aren't I good enough for you?' he said. 'Is there something wrong with me that you want to keep dashing back there?' She suggested they needn't *stay* with Paul, they could go to a boarding-house near by. Take walks in the mountains. 'I only walk when my car don't go,' he said. 'Walking's humiliating.'

Norman hit her from time to time. No man had ever hit her before, apart from Sal, once, when he believed standards were being seriously breached. Norman didn't have standards, he just had a temper.

He kept disappearing for days at a time, leaving her alone in the studio. When he returned, he'd just say, 'I had to be on my own.' Or, 'I had to go to London on business.' Leah asked him why he couldn't give her some notice beforehand, but he had an explanation for that as well. 'I get an offer and I got to act on it. I don't have time to leg it home and tell you the good news.'

One time he didn't show for fifteen days. He had left her and the kid without money for coal, electricity or anything else. She wrapped Jake in blankets and herself in a rug. It was the middle of a cold winter. The temperature outside was five below zero and catching up inside. After two weeks, she left Jake with a neighbour and went across town to that house of his in Picton. She walked for an hour through slush-filled streets on a thin line of hope.

A woman opened the door and a blast of warm air hit her in the face. Leah's head was down in her boots. Slowly she worked her way up from the woman's slippered feet, shaved legs, cleavage. She had a kind of compressed beauty channelled through thin lips and bleached eyes.

'I'm looking for your landlord,' Leah told her.

'Norman ain't my landlord, he's my fuckin' husband!' the woman expostulated.

The walk to Picton had been cold, but nothing chilled her more than that nugget of information. She turned around and ran into her own frozen breath. Somewhere on the road she lost one of her shoes. As she limped home cars slowed up and followed her along the kerb, crunching snow under the tyres. Drivers leant out of the windows and licked their lips.

Leah was in the middle of packing when Norman appeared. 'I was too afraid to tell you, Leah. I was scared you'd leave me. But I don't love her. I wanted to leave her years ago. I felt sorry for her. That was my big mistake, marrying someone for pity.' Leah's resolve was smothered by his repentance. An hour later they were in bed, his head in her lap. It looked like there was no life for her without Norman.

Norman behaved like an Augustinian friar for the next few weeks. He was attentive and generous, out during the day but home by five sharp. Leah would make his supper and he'd eat with his plate on his knee. One night he came back wearing a bow-tie, carrying a bunch of orchids. Parked outside was an invulnerable-looking Ford Pilot V8 in deep midnight-blue. 'Now we can go to see your brother,' he promised. He waved a contract in front of her face prodding at a figure of seventeen thousand pounds. Before she had a chance to study the small print, he'd folded it and put it back in his pocket. He carried her in his arms like a bride to the bedroom. He wanted to celebrate in the ancient way. She obliged, trying to be what he wanted her to be.

Norman's seventeen G contract 'fell through' before the end of the month. The car went back to the showroom and the sheriffs returned. This time they were served with an eviction notice. Leah asked about the house in Picton and he swore that he'd seen a lawyer about getting a divorce. Until then, he couldn't get into the house. He never produced any evidence of this. He started punching holes in doors, smashing furniture. She looked at his bloodied hands, those weapons of offence, and could no longer bear him touching her. She realized that, all in all, life with Norman was inferior to the life she'd left behind. No man in Sharon supported two wives. She'd moved to an ungodly place. Paul had tried to warn her about the world outside the family corral, beyond Sharon on the other side of the mountains, where the temptation was for men to get what they could out of it without paying their dues. That was Norman's philosophy in a nutshell.

Leah decided to leave Norman the next day, while he was out running his spectral business. The morning came slowly. At nine Norman went to work in a the heap of shit he called an automobile, the rust falling off as he drove away. Leah was halfway through packing when he made an unprecedented return to the studio. He saw the suitcases on the floor. With his hands he tore out a floor-board and swung it at her. She reached over into the sink and grabbed an empty milk bottle by its neck and broke it over his head. He looked at her like a big old stupid bear surprised by someone's headlights in the Michigan woods. Then he fell over. Leah picked up the bag she had finished packing and held Jake by the other hand. Norman was flat out on the threadbare rug. She made for the door. Norman shook the air with a piercing shriek. She told him she was going. Norman held his head, the blood pouring through his fingers. 'Let's talk it over,' he said.

'We can talk but I doubt whether we'll actually say anything. We haven't had a conversation since the day we met. Nothing real. If we did, you'd fall apart . . .

'And then I just left. I walked out through the rain and the first thing I did was trip over a paving-stone. But what a long road it has been to this door, Aaron.'

And then the door opens and a tall Negro in dungarees

steps inside, ten years her junior, his face sleek with diesel oil. And what a piece of art he is, with a finely sculpted head and long fingers that he places on Leah's shoulder. Black is such a complex colour. In those tumbling few seconds I take account of his skin, swimming through a spectrum of blue, purple, brown, olive, pink. All the foreign travel a lover could handle.

She introduces us to the boyfriend. He has one of those African names that I can never remember.

'Is that a Wolof name?' I ask.

'Mandinka. You came close.'

'You born in West Africa?'

'Me dad was. I'm a scouser.'

As though to impress that her guy is more than a docker, Leah tells us he's a jazz trumpeter.

'Yeah?'

'Blues and jazz. Em, do you play?'

'You heard of the bassist Percy Heath?'

'Sure. And his brother Jimmy.'

'I saw him once. The Modern Jazz Quartet. I saw them play in the Showboat in Philly.'

'You've *seen* the Modern Jazz Quartet?' He leans out of shadow towards me.

'Also John Coltrane. *He* doesn't actually come from Philly, but he settled there, from the Carolinas. And Archie Shepp, I've seen him too. With Henry Grimes and Bobby Timmons. Yeah, I was really interested in Archie Shepp, a Negro opposed to racial integration. Said he didn't want to lose his origins.' Maybe Shepp was being ironic, who knows for sure. That's what blues is, after all, an ironic expression. Africans invented irony to help them through slavery. The white man came along and stole it off them along with everything else. 'Shepp said once that the Negro knows how to reaffirm life because he knows what it's like to die prematurely.'

I watch all this information work its way through his pretty head. He sifts through my remarks about Archie Shepp, scouting for innuendo. He doesn't know what to make of me. But all I'm doing is talking through my nervousness. I happen to share Leah's tastes in men.

'Remember when we danced to Count Basie in the chapel?' Leah asks.

I smile briefly in acknowledgement, but no more than that. For all I know, the boyfriend is as possessive as his predecessor, Norman.

The sidemen in his band come over later, three scouse Negroes who sit in the parlour together, completely indifferent to the neighbours' cats and kids crawling all over them as they listen to records revolving on the turntable: Louis Jordan, Lionel Hampton, Jimmy Witherspoon, Joe Turner, Peggy Lee, Ornette Coleman, Dizzy Gillespie . . . Every one an American.

That's something else I've noticed since being back. Jazz, blues, the movies, have followed me over. It's an infestation; colonization in the shape of entertainment. We are *all* Americans now.

The boys cue in to how Americans do it, commenting with a howl or a whistle when they hear their own particular instrument stretched to the limits. They take swigs of sweet red wine between tracks. Leah keeps putting her head out of the tiny kitchen, where she is making a chicken stew dressed in peanut sauce, filling the room with smells of spices. She holds out an empty glass and pleads with the boys, 'How about another drink for the cook?'

All this activity dwarfs Glanmor, watching from his editorial position in a threadbare lounger, his hands gripping its greasy arms. But he will have to get used to it. Black absorbs white, not the other way round.

Ella Fitzgerald takes her turn on the gramophone. Her voice is clean and direct. Nothing adjectival. Leah forgets herself momentarily and sings along with the track, 'T'aint Nobody's Business but My Own'. This *is* a revelation. Leah has a beautiful voice. I roll back my shoulders and clutch the back of my chair with both hands. 'When did you learn to sing?'

She comes out of the kitchen fanning her hot face with an album sleeve. 'I never did learn.'

'Your father was a good singer.'

'No one ever told me I could sing because no one was bothered to find out. They didn't have much use for girls in male voice choirs.'

'She sings with our band,' the boyfriend announces proudly.

'Get out of here!'

'Yeah, man, she's the new Mildred Bailey,' he says. 'The rockin'-chair lady.'

I wave my arm at him. 'Peggy Lee! She's the role model for her. A blonde in a beret whose eyebrows used to furrow on a blue note.'

'The boys told me my name had to go. No shouter was ever called Leah Gravano. So I renamed myself Sharon Wales.'

'Sharon Wales is good,' I say.

'Sharon Wales and Synchro System.'

After supper the band close ranks and put in an hour's rehearsal for a gig they are to play at midnight, in a club called the Jacaranda. I watch them assemble behind their instruments. The stocky drummer with a pudding basin of curls dyed blond keeps time with his sticks on the edge of the table. The piano player, shiny as an eel, is slow and ponderous, his manner cool and detached. He wears glasses that are non-magnified, a conceit I later discover is in reverence to Thelonious Monk.

I can't stop myself from smiling, amazed at the community that Leah has found. It's so . . . *adventurous*.

'What key do you want to sing in, like?' the boyfriend asks.

'Just play and let me feel my way round.'

'You got to sing in one key or another.'

'I'm all right. Just get on with it. I'll come in when I'm ready.'

The piano player tries to lead her in to one of the songs by pressing down gently on the piano keys until the boyfriend stops him. 'We know you used to play in church, Sly, but you've put that behind you now.'

'I'm trying to go easy on that girl. Besides, I don't know what they're playing in church these days. But I played jazz, I know tha'.'

'Then play it now, all right?'

He chops the keys, using his elbows, jabbing and teasing out a riot of notes.

'Stop, stop, stop, stop.'

'I'm trying to play like Monk.'

'No one can play like Monk. Stop fooling around.'

He makes another start. This time he seems to pick the notes out of the air, and the melody is suddenly all around us, practically visible.

Leah starts off on shaky ground. She sings half a song, following the piano, when the boyfriend suddenly shouts out, 'Hang about. I wanna waird wi' you.'

'What now?'

'You're singing folk music.'

'Sorry.' She tries not to sound offended. 'But I also used to sing in church.'

She tries again but he stops her once more. 'You got to lift the melody from the ground beat, phrase against the beat. Don't bring your notes down on it. You understand what I'm saying.'

'No, I don't.'

'Don't come down on the beat. Work against the beat. I don't know how else to say it.'

'Let her listen to some a' Billie Holiday.'

'I can't sing like her,' Leah says. 'A white woman can't sing the blues.'

'There's enough blues in your life.'

'I can't do her kind of thing. Her singing follows the patterns of American speech.'

'You got good phrasing. In a while with the groundwork down, it'll come easier,' the boyfriend says. 'Right, fellas, let's run it down. Billy . . . start down there with the shading.'

We leave the house round quarter to eleven. On this mild December night people have left their homes to the cockroaches and congregate in the street, cooking over charcoal, eating as a big team. All along Upper Pitt Street fires glow in the night. The air smells of curry and fried fish. We pass groups playing sabar drums, goje fiddle, finger pianos, koras. Two leathery old guys play bottleneck guitar and banjo on the stone steps of number 215, which, I am informed, is a former mayor's house, and sing American songs learnt off indestructible 78s that arrived on the Cunard Yanks. Further on up the road a group of kids are singing the *a cappella* gospel song 'Oh Mary, Don't You Weep'. It's a fusion of African and American music – one time the same thing – the former imitating the horse-cart rhythm, big birds whistling; the latter a babylonian cacophony of sound.

I feel much happier than I did a couple of hours ago. All this music makes gradual sense of Leah's decision to live here. And the docks simmer nearby, offering a never-ending supply of labour for these men. 'I enjoy singing more than I ever thought I would,' she tells me, her arm linked in mine. 'To discover that I had a talent so late in the day is an extraordinary thing. I'm earning a living at last, Aaron. It never occurred to me that my voice was worth paying to listen to.'

With the boyfriend and Glanmor leading, we stroll along Parliament Street, one long conduit of clubs: Joker, Tudor, Woodbine, Sierra Leone, Blue Angel, Palm Grove, Berkeley's Arms, Harry's Shore, Rosie's, Brass Bar. Leah has played them all.

The Jacaranda Club is in the basement of an imposing colonial house, a former slave trader's home, now thoroughly rinsed with the blues. From the outside there is no way of telling what is happening inside. The boyfriend knocks on the door and a peephole slides open. Brown eyes check us over. 'Come on, Uncle, let us in,' he shouts.

'Uncle', also known as the Dutchman, opens the door and I follow them down a stairwell into a room packed with three-hundred-odd people, Africans and scousers and Negro servicemen from USAF Burtonwood, who have spiced up their khaki uniforms with little personal touches: white scarves around their necks, Hawaiian shirts under their service dress. I think it seems a no-trouble place until I notice some of the GIs have brought their side-arms with them.

While Synchro System warm up the audience, I sit at the

Dutchman's table. He casually places his revolver on the tabletop, to serve as a reminder to the band that he is paying them good money to play at a topnotch level. Leah sits on the other side of the Dutchman, wearing a black velvet dress with silver embroidered bodice and fake diamond necklace, reducing the Dutchman to a pussy cat. 'You look good,' he says.

'But I feel like shit.'

The Dutchman sees the worry all over her face. 'So listen, eh, your boy's a fabulous musician. The governor. You goin' to be all right.'

The boyfriend places the horn carefully to his lips, his Adam's apple going up and down once, before inhaling a lungful of air clogged with tobacco smoke, sending it back as clean notes. He gives us something carefully considered, something quiet at first, a peaceful arrangement of notes, with a breath of tone here, a sweep of phrasing there, playing dance-floor licks over different rhythms. Impersonating an air-raid siren, he links the piano and sax solos that follow. Two drunken GIs dance cheek to cheek on the floor.

My eyes are full of tears from the smoke in the room. 'He sure can preach with that trumpet,' I say, but Leah is too nervous to reply, she has just been given her cue. She walks on stage and stands in front of the mike. The first bars she sings are so limp that she pulls out, down the back of the stage. The boyfriend wanders over very casually as if he has nothing special on and whispers in her ear. He raises the horn to his mouth and charges through the chords into a solo that lasts a long time, before reintroducing a fragment of the theme. Suddenly it is time for Leah to go again. She comes in to the mike and chokes for a few awkward seconds. Then she sings 'Easy in Love' with the trumpet enhancing her phrasing in a tender way, like Lester Young behind Billie Holiday, or Paul Quinichette behind Dinah Washington.

She continues the set with 'Fine and Mellow', 'You're Too Marvellous', 'Please Don't Talk about Me'. I measure the effect her voice has on the audience, drunk on Juanita and Esta Lina red wine, dancing into a trance, into a state of possession, as if at a revivalist meeting. She sings like an African, with no vibrato, no shake at all, sermonizing like Paul, in the way he used to make us all feel good. Except this isn't politics, it is sensual entertainment. Her voice flows through my body, a warm fusion of her ego, the band playing around her. I wonder if Paul ever felt the same. But I figure in his case the messenger was subsumed by the message.

With each song her confidence grows. She imposes her own licks, decorating the melody in some of her own colours. She floats out on the low notes a tender grief that makes me feel, I don't know, nostalgic, even though I'm not sure what for. She switches from a thumping earthiness to ethereal chapel sentiment.

In the middle of 'What a Little Moonlight Can Do' some white guys in the club start shouting, 'Rock 'n' roll! Rock 'n' roll!' Leah dries up as the Dutchman climbs out of his seat, taking his revolver with him. 'I'll fuckin' brain 'em.'

He returns a few moments later after ejecting the hecklers from the club. But Leah's morale has been badly shaken and she struggles to get to the end of the set.

She comes off the stage and lays her head on my shoulder. The sidemen all follow her to the table to insist that it isn't her fault. The boys explain to me what's happening here. A couple of years ago the movie *Love Me Tender* was shown at the Empire. Five hundred teenagers came out of the picture house and got chased down the street by the police. After seeing the film they just wanted to break something. Since then rock and roll has been gaining momentum in Liverpool like a rampant plague, especially on the northside, where white folk were going around rabid and expectant. They wanted to hear establishment deprecation in the songs. Scouse bands were performing homemade covers of Chuck Berry, Fats Domino, Little Richard, Muddy Waters. A & R scouts were drifting into Liverpool with their cheque books open, but no one ever saw them on the southside. They wanted to record the new R & B sound but with white bands playing it. Some of the clubs on the southside were bowing to the pressure and letting in rock and roll outfits. The Dutchman was one of the few to stay 'pure' . It all seems to point to the end of something. A new music culture was dawning and that made Sharon Wales and Synchro System nervous.

'Rock 'n' roll is nothing but warmed-over blues, anyhow.'

'Every bloody teenager wants to play a guitar now. They keep coming over the southside for tuition.'

'I've been teaching this bloke for a month,' the boyfriend tells me.

'That twat? He couldn't tell a quarter note from a beer glass,' Leah says. 'When he first came to our house he took one look at me and asked if I was being held hostage.'

'How long will this thing last? . . . Anybody know?' I ask.

'It'll blow over.'

'It's been two, three years now. Long time for any breeze.'

'Wait till the kids get older, man. Rock 'n' roll's for adolescents.'

'Johnnie whatsisname – you know, the Nabob of Sob,

243

Prince of Wails – he gets the kids going by falling on his knees with *emotion*. But drop him over the jetty in Queens, he still wouldn't sing in tune.'

'When I went over there a year ago, people on Scotland Road started touching my skin and pulling my hair. It makes you wonder, doesn't it, when you think the Negro's been in Liverpool two hundred years.'

I sip my wine listening to them. They consider the Bill Haley phenomenon. 'How long can a family man with a paunch keep the torch lit?' And Elvis Presley, who worries the hell out of them. His latest record had just come in on a Cunard and the bluesmen got a preview of what the damage was likely to be. 'How can you compete with an onslaught like that? He's merging everything . . . country, blues . . .'

'Elvis is just another white boy who thinks he's black. Is he better than Chuck Berry? Little Richard? Joe Turner? Better than Otis Blackwell who wrote "Love Me Tender" and "Great Balls of Fire"? Man, it's *vulgar*, that's what it is. Blues is our music. They're plunderers. Fucking A-one slavers.'

A couple of bloated faces appear in front of us, hovering like melons on poles. One leans into the drummer and slurs, 'Go and play rock 'n' roll, nigger,' before dissolving into the crowd, making for the exit.

The drummer wants to follow them into the street 'to sort 'em out'.

'You do that,' the boyfriend reasons, 'and you'll be chasing white boys down the street every day of your life.'

But the drummer gets up and goes anyway. Leah follows the boyfriend who follows the drummer up from the table and I follow her, ordering Glanmor to stay put. Outside the club, the two white guys stretch their legs, leaning against a car. The drummer walks straight over to them and I hear him say, 'I want a waird wi' you about your fucking language,' before he is bundled against the hood.

'Oh shit, here we go . . .' The boyfriend steps in. In my haste to back him up my foot slips in a pool of oily water. Skidding, parting the water into a multi-coloured film, I see one of the thugs draw back his arm, about to throw a punch. Leah has seen it too and shouts. The boyfriend ducks and the punch goes wide. He pushes the thug as he is off balance. He falls into a similar pool of the oily water that is giving me so much trouble. I feel old suddenly, too old for this game. Meanwhile Leah has wedged herself between all these big male bodies and holds her ground, silently negotiating with them. Nobody could hit anyone without hitting her first. They stand around her with their fists hanging by their sides. The Negroes fade into the night and the moon shines on her.

The band go up to play their second set. Leah stands on the sidelines waiting to go on. The band play her cue but she can't seem to move. Then someone gives her a shove in the back and she has to run to stop herself falling. Suddenly there's the mike in front of her nose. Her legs begin to wobble. She is all over the place. I hear the downbeat begin and Leah starts in. Her voice sounds like a slipper skidding on a carpet, a broom falling over in a cupboard, until she gets it roped in and begins enjoying herself, as she did in the first set, belting out four ripe songs on the hoof.

I dance in Glanmor's arms in the shadows and give a passing thought to the deacons in Sharon who equated dancing with manslaughter in their list of heinous crimes. They who chased Leah out of Sharon and did her a favour. Leah enjoys the life here. She wants to be out every night, preaching with that voice. If she had stayed in Sharon, as her brother wished, she'd have been *in* every night. She wasn't ready for that kind of life then and she isn't now.

The guys thrown out earlier have somehow managed to get back in. This time they have brought their friends, dozens of them, who keep chanting 'rock 'n' roll, rock 'n' roll'. In the corner of the room I see a man quietly go down with a glass broken in his face. The spotlight glances off a revolving spherical mirror in the ceiling and dashes all that happens with pebbles of light. It lends the brutality the dreamlike quality of a fairy tale. But not for long. Fights start breaking out all over the club. The Dutchman piles in. Synchro System stop playing, the spherical mirror stops revolving, the house lights go up and suddenly it is all too real. A woman screams. I look behind at a sea of white faces. I see a chair fly and catch the boyfriend on the shoulder. A second chair chops the piano keys before flipping on to the stage. Dozens of missiles and glasses and pieces of furniture follow, flying out of darkness into light as they cross the spots, like a flock of crazed buzzards seeking eyes to pluck. White hooligans with knuckledusters and steel chains pour through the upturned tables.

I steer Glanmor out of the club, desiring to stay in one piece for our embarkation later in the morning. We emerge into the deep blue kinetic of Toxteth around three and relieve ourselves against a stucco wall. The city unscrolls in front of us, emitting small clunks and chirrups and the crackling gunfire of 9mm weapons as GIs head home to their base. Beyond the gasp and leap of the docks, the River Mersey is strafed by bands of yellow, green, red, white light. It is a stage set for emotional departures as blazing cathedrals set keel for open sea, transporting souls from their fixed coordinates to un-native soil, from this day to the next, from generation to generation.

Author's Note

The original sources for *An Interference of Light* were my father's own deeply ethical reminiscences of an erstwhile community of slate quarrymen, among whom his parents were marginally involved until their exodus to South Wales at the turn of the century. I have managed to re-invent just about everything that I was told or have since read, but the moral climate of the book, I think, is close to the historical truth. My father never got to read my novel, but his fingerprints are all over it.

Pinkerton's National Detective Agency, so far as anyone knows, did not extend its industrial spy system to distressed businessmen in Great Britain, although it is on record that they assisted Scotland Yard during Edward VII's coronation in keeping 'anarchists and dangerous criminals' away from the event. While researching Pinkerton's industrial activities I found most useful James D. Horan's general history of the agency, published in 1979, and Allan Pinkerton's 1897 hagiography of one of his operatives, James McParlan, who remained undercover in the Pennsylvania coal fields for two years.

I have also drawn from various accounts of the Penrhyn slate quarry lockout of 1900–1903 among which R. Merfyn Jones's *The North Wales Quarrymen* was indispensable.

The quoted extracts from Hiram Bingham's diary, recording his Calvinist mission to the Hawaiian Islands in 1820, I came across in the *New Yorker* a few years ago, as part of an article on surfing by William Finnegan.